MATHEMATICS OF CONFLICT

NORTH-HOLLAND SYSTEMS AND CONTROL SERIES
VOLUME 6

NORTH-HOLLAND
AMSTERDAM • NEW YORK • OXFORD

MATHEMATICS OF CONFLICT

edited by

MARTIN SHUBIK
Cowles Foundation for Research in Economics
Yale University
New Haven, Connecticut, U.S.A.

N·H
P·C

1983

NORTH-HOLLAND
AMSTERDAM • NEW YORK • OXFORD

ISBN: 0 444 86678 7

Publishers:
ELSEVIER SCIENCE PUBLISHERS B.V.
P.O. Box 1991
1000 BZ Amsterdam
The Netherlands

Sole distributors for the U.S.A. and Canada:
ELSEVIER SCIENCE PUBLISHING COMPANY, INC.
52 Vanderbilt Avenue
New York, N.Y. 10017
U.S.A.

Library of Congress Cataloging in Publication Data
Main entry under title:

Mathematics of conflict.

(North-Holland systems and control series ; v. 6)
Includes bibliographies.
1. War--Mathematical models--Addresses, essays, lectures. 2. War games--Addresses, essays, lectures. I. Shubik, Martin. II. Series.
U21.2.M364 1983 355'.02'0151 83-8118
ISBN 0-444-86678-7

PRINTED IN THE NETHERLANDS

INTRODUCTION

If theory caused a more critical study of war it would have achieved its purpose.
—Clausewitz

These papers are primarily the outcome of a conference on the theory of conflict held at Seven Springs, Mt. Kisco, New York in November, 1979 by the Project Center for Competitive and Conflict Systems Research of the Cowles Foundation at Yale University sponsored by the Office of Naval Research.

The concept of the use of calculations in war dates back to the fifth century B.C. when Sun Tzu, the father of the theory of strategy, observed "Now the general who wins a battle makes many calculations in his temple ere the battle is fought. The general who loses a battle makes but few calculations beforehand." Since that time the division of the cultures has been present in the students of conflict. The diplomats and softer social scientists and many generals question the worth of formal models and mathematics when dealing with strategy. Certain geometrical manuevres have made their way into tactics and naval strategy such as "crossing the T"; but the preoccupation with either geometry or numbers seems to bespeak of a spurious accuracy and irrelevant attachment to formalism in a world where the imponderables dominate.

There is no question that there is a mathematics of conflict, be it in the writings of Sun Tzu, Jomini, or Mahan, even though no direct mathematics appears in their writings. The concept of the "fleet in being," the aphorism that "Force does not exist for mobility, but mobility for force" (Mahan, 1890) have the chasteness of concept and precision of thought that serves as a sound basis for subsequent mathematisation.

The growth of operations research in and since World War II has seen a proliferation of mathematical models of conflict. The Lanchester equations, the work of Lewis Richardson, and finally the development of the theory of games provide ample modern examples of formal mathematical models of conflict. The mere construction of mathematical models is not the same as the construction of useful or insightful models. "I can call spirits from the vasty deep. Why, so can I, or so can any man; but will they come when you do call for them?" (Shakespeare, Henry IV, Part I). The production of mathematical models poses no barrier to science or thought. The production of relevant, insightful pertinent mathematical models poses a problem of a different order of magnitude.

The purpose of this volume is to suggest that the formal modeling of conflict has become both more sophisticated and promising. A blend of mathematics, insight, verbal description and perception of operational relevance is evolving. Essays and numbers; history and equations must be melded to produce insights which could not be obtained without the intermix of quantification and qualification.

The six essays which comprise this study vary considerably from "soft and squishy" to hard and precise. The first two by Shubik and Strauch are addressed basically to the problems of modeling. Shubik considers the uses of the theory of games in the provision of a language for the analysis of strategy and conflict in general. He stresses the important distinctions made by Clausewitz between the science of war and the investigation of war as a social process. Strauch is concerned in general with policy analysis and quantitative methodology and its supporting theory. He offers a critical assessment of the usefulness of quantitative methodology in policy analysis and stresses the difficulties with the implicit assumption often made, that the methodology of the model builders and mathematicians is to be based upon well-defined problems; whereas in reality in the actual process that leads to the sponsorship of the analysis the parties concerned may not know what they want or if they do they may be using the analysis for purely partisan or advocacy purposes. Both Shubik and Strauch lay emphasis upon problems of perception and interpretation in model building.

The paper by Weiss was written almost thirty years ago, yet it remains one of the clearest and most persuasive exemplars of why and how mathematical models can be of considerable value in the investigation of tactical problems. In particular, the discussion touches upon the use of weapons, information, terrain, combat effectiveness criteria and communications stressing in each instance where the possibilities exist for the use of historical information from actual battles and where hard measures can be obtained from tests. Throughout the discourse a balance is maintained among theory, mathematics, fact and operational questions.

Kimeldorf presents a detailed survey of the developments in the mathematical theory of duels. The theory of duels is a highly technically developed branch of two-person zero sum game theory with direct application to weapons evaluation problems. The essential model involves two individuals firing at each other with a host of factors determining kill probabilities such as range, frequency of fire, availability of ammunition, and so forth. In contrast with the Lanchester equations noted here by Karr, Weiss and Shubik, dueling theory concentrates upon the "1 on 1" aspects of combat. Thus tank versus tank or aircraft versus aircraft pose problems of concern. Communication variations have also been considered and the theory has been developed for both noisy and silent duels. A silent duel is one where the opponent is not aware that you have fired unless he is hit. Kimeldorf presents a considerable bibliography together with his survey.

Most of the literature on dueling is theoretical and many technically difficult and unsolved problems exist. Slight changes in problem formulation may make a considerable difference to the model. For example, in 1 on 1 fighter plane combat the introduction of a "trigger freeze" delay between when the pilot has decided to fire and when firing takes place, or a small blind spot time when visibility might be blocked by the trail left by the weapons fired, may considerably change both the difficulty of the mathematics and the nature of the results. Another modeling problem with duels involves the meaning of the payoff and the definition of the end of the game. In general the zero sum assumption is reasonably good, but not in all applications.

Duels concentrate on "rational models" of 1 on 1 fighting. Lanchester equations on many versus many behavioral models of fighting. There is a gap in the mathematical literature concerning the conscious selection of targets in many on many situations. A fanciful illustration of this is given by several articles on "Truels," but these are more for demonstration of paradoxes than for calculation in applied battle models (see Shubik, 1954, 1982).

The paper by Karr provides an extensive overview of the major variations of the Lanchester model together with a considerable bibliography. It also provides a discussion of several applied models used for attrition computations in theater-level combat. The encyclopedic work of Taylor (1981) contains both more detail and references; but this far briefer coverage is nevertheless in sufficient depth to illustrate the scope of the models, to provide a qualitative understanding and an appreciation for the problems in application and the utilization of simulations. The choice of specific models was based upon the availability of documentation and use in studies and analyses.

Karr is particularly careful in distinguishing interpretations of the basic equations and the underlying assumptions that some have chosen to make.

Ho and Oldster provide a self-contained introduction to differential games treating for the most part two-person zero sum game models. Extensive references to the literature are provided. Although it is noted that the authors have not been able to provide full coverage of the proliferating literature on differential games in the Soviet Union. Among the types of problems considered are search and pursuit games several of which appear under the frivolous titles of the homocidal chauffer, the body in the lake, and the suicidal pedestrian. Specific military problems of relevance include convoy search, surveillance problems and dogfights. However, the mathematics is in general extremely complex and the stripped down models are overly simple. If one adds any geographical features the problem may change considerably.

It is noted that Richardson's model of armament build up between two countries can be recast as a non-zero sum version of a differential game. Unfortunately the concept of solution (see Shubik's paper in this volume) for non-zero sum games is not agreed upon. The Nash noncooperative equilibrium, frequently used, is subject to considerable proliferation, does not in general give a unique value and must be interpreted with extreme care.

The authors' assessment of differential game theory is that its value lies in the way a group of problems are considered rather than in its current ability to provide specific solutions to applied problems.

It is of interest to note that control theory and its near relative, differential game theory, have considerable sponsorship in the research institutes of the Soviet Union; yet the gap between application and theory is large. Furthermore the problems faced in application as judged by the relevance and robustness of the modeling assumptions are enormous.

One possible explanation for this phenomenon might lie in the sociology of science and applied science rather than within science itself. The questioning of basic assumptions must take place in the context of the socio-political and bureaucratic life of the country. By its very nature it has to be a dangerous occupation which is bound to upset some generals, institute directors, senior professors, or politicians almost all of the time. A way to avoid this is by developing a priesthood in such a manner that no matter who is in power, nor what he actually thinks of the priesthood, the Tzar will still pay the Patriarch for his symbolic blessing be it with differential games, duels, Lanchester equations, catastrophe theory or large scale simulations. The "can do" lesser priests will supply the troops with ad hoc blessings in the form of simulations, operational games, assessment models, and other quick-fix number producers. These must have sufficient plausibility that as long as the basic honor, beliefs and understanding of the commanders are not too badly challenged they will provide for the celebration of the mathematical high masses.

This volume is presented in the firm belief that an important way to promote the chances for peace is to better understand the social process of war. The models of conflict are innately mathematical. The modeling of conflict situations is part of a social process where policy, science, engineering, and mathematics are intermixed and welded by the institutions of society.

REFERENCES

[1] Mahan, A. T., The Influence of Sea Power upon History 1660-1783 (Sampson, Low, Marston & Co. Ltd., London, 1890).

[2] Shubik, M., Readings in Game Theory and Political Behavior (Doubleday, New York, 1954).

[3] Shubik, M., Game Theory in the Social Sciences (Massachusetts Institute of Technology Press, Cambridge, 1982).

[4] Taylor, J. G., Lanchester-type Models of Warfare (Naval Postgraduate School, Monterey, California, 1981).

TABLE OF CONTENTS

MATHEMATICS OF CONFLICT, Martin Shubik (ed.)

GAME THEORY: THE LANGUAGE OF STRATEGY?

M. Shubik

Cowles Foundation for Research in Economics
Yale University
New Haven, Connecticut
U.S.A.

> We say therefore War belongs not to the province of Arts and Sciences, but to the province of social life. It is a conflict of great interests which is settled by bloodshed, and only in that is it different from others. It would be better, instead of comparing it with any Art, to liken it to business competition, which is also a conflict of human interests and activities; and it is still more like State policy, which again, on its part, may be looked upon as a kind of business competition on a great scale. Besides, State policy is the womb in which War is developed, in which its outlines lie hidden in a rudimentary state, like the qualities of living creatures in their germs.
>
> Whether such a conflict of the living, as takes place and is settled in War, is subject to general laws, and whether these are capable of indicating a useful line of action, will be partly investigated in this book; but so much is evident in itself, that this, like every other subject which does not surpass our powers of understanding, may be lighted up, and be made more or less plain in its inner relations by an inquiring mind, and that alone is sufficient to realize the idea of a theory.
>
> Clausewitz

1. THE SOCIAL PROCESS AND SCIENCE OF WAR

1.1. The Social Process of War

The social process of arming a nation, preparing for the possibility of conflict; threatening and negotiating using the potential of military power to achieve political goals while avoiding direct conflict; is far from the same as the study of war or diplomacy or the giving of operational or moral advice.

The analogy of Clausewitz can be extended. The business of war stands in the same relationship with the science of war as does the business of the economy with the science of economics.

Many individuals in a complex social network are embroiled in the business of war. Politicians, diplomats, generals, scientists, industrialists, advisors, budget bureaucrats, lawyers all are engaged in a complex social process involving advocacy, cooperation, adversary struggles, science, intuition, goal formation, moral introspection and value formation.

Perception, responsibility, fact, volition, belief, power and desire are blended by the many forces at work to produce an overall perception and policy. This policy, to many individuals be they senators, generals or scientists or the public, may be far from their own beliefs, yet is partially the result of their inputs into the process.

All individuals, even remotely associated with the act of war have a moral involvement and responsibility at large. The politicians and statesmen and other political, social, religious and intellectual leaders of a society have a responsibility for the overall timbre of policy. The generals and the military establishment as a whole have the prime responsibility of executing policy in an efficient manner and for pointing out whatever weaknesses or flaws which exist in policy in terms of the feasibility and efficiency of execution. The historians, physical scientists, social scientists, engineers, operations researchers and other advisors have a scientific and metascientific responsibility. At the least they must verify facts and muster more or less objective opinions about a host of questions. But beyond the bare minima, they must also challenge the relevance and the precision (or lack of it) in the specification of the questions being asked. The ability to deepen understanding and clarify insight depends upon the knowledge, analysis and artistic insight which characterize scientific enquiry.

Knowledge, responsibility, strength of purpose and social perception are not all required for good scientific understanding. They are, however, called for in framing good overall policy. The adviser (Goldhamer, 1978) plays many roles, but taking direct responsibility for blending administrative, political, social and scientific inputs is not among them.

The historian, social scientist and other advisers are cast in an anomolous role. They are expected to provide objective judgment yet to advise and counsel on social process. If for no other reason than to establish better communication between those responsible for evolving and executing policy and those responsible for understanding the scientific implications, the development of a language of strategy is called for.

1.2. The Art and Science of Studying War

It is suggested here that the study of war is a social science, much like economics of political science. Like all of the other social sciences it has been subjected to a considerable debate concerning the role of mathematics. Furthermore, in all of the social sciences there has been a tendency to equate the mathematical formulation with the scientific content. To a great extent this has come about by a failure to recognize three rather than two important distinctions in the utilization of science in the processes of war. Specifically there are (1) those responsible for policy and execution, (2) those responsible for modeling and setting up problems to be analyzed and (3) those responsible for analysis.

Perception and cognition are not the same as analysis. The mathematical model may be precise and deep. The mathematical analysis may be elegant and difficult. But no amount of mathematics in and of itself spells out good science in general, or military science in particular. The good scientist requires the ability to perceive and formulate the relevant problem prior to the analysis. The good scientist when operating in an environment where he has to both communicate with his colleagues and bear witness to the social worth of their studies must transcend many boundaries in an effort to promote precision of thought and good analysis.

The different roles per se in the development of the art and science of war are discussed no further. Attention is focused upon one specific aspect. That is the need for a language to aid in model building, analysis and communication in the study of conflict. It is suggested here that the theory of games may provide this language.

2. THE LANGUAGES OF GAME THEORY

The very title of game theory from its inception was a misnomer. It was a misnomer caused, in essence, by the clash of the mathematical and non-mathematical cultures. Von Neumann can be credited with the title game theory derived basically by analogies with chess and poker which were the sources he used in the development of the notation and many of the underlying concepts of the theory of two person constant sum games. The well defined formal structure of games such as chess was an obvious starting point from which a precise descriptive language, concise enough to facilitate analysis could be invented.

Unfortunately, language in scientific usage may easily clash with language in social use. To the diplomat or general the word "game" may be synonymous with "fun and games," with entertainment and frivolity. By attaching the word game to a military exercise or analysis the social communication may be made harder. The work of the two Von Reisswitz, father and son, were first regarded as games, all be it war games. Their adoption and utilization were by no means guaranteed until the chief of staff of the Prussian army, Von Meuffling, exclaimed, "It's not a game at all, it's a training for war; I shall recommend it most emphatically to the whole army."

It is my contention that in actuality there is no single theory of games but there are several different topics the most important of which is a powerful, formal language for the description and analysis of strategic and tactical situations. There are also several theories of behavior which contain normative prescriptions as to how one should behave or behavioral descriptions of individual action in strategic or tactical encounters.

The language of game theory provides a way to outline the anatomy of cross purpose optimization. In other words, it provides a precise description for situations in which there are many actors with independent goals and only partial control over their environment.

In this section the formal structure of the description of strategic situations is sketched. In Section 3 the various concepts of solution are discussed. In Section 4 both the formal language and the solution concepts are recast in terms of military strategy and diplomacy. Stress is laid upon the problems of modeling, perception, and interpretation. Study and analysis is different from sponsorship and utilization. It is precisely this difference that spells out the distinctions between the science and art of war and war as a social process.

2.1. Scope, Detail, and the Rules of the Game

The formal methods of Von Neumann and Morgenstern utilized to describe strategic situations provide three different descriptions of multi-person cross purposes optimization, each based upon different levels of scope and detail and different purposes for the use of the models. The three descriptions are known as: 1) the extensive form of a game; 2) the normal or strategic form of a game; and 3) the coalitional or characteristic function form. The first deals primarily with process, information, and details of behavior; the second is designed for the analysis of strategic policy; and the third is to aid in negotiation and diplomacy and in other situations in which there is the possibility of cooperative behavior resulting in mutual benefit. Most international dealings (and for that matter most dealings among individuals) can be characterized as giving rise to mixed motivations. There are both elements of potential conflict and co-operation. Each may want as large a slice of the cake as he can claim, but by cooperation the cake itself can be made larger. Situations involving pure opposition of interests are relatively rare and occur, for the most part, at the level of tactics.

2.2. Process and the Extensive Form of a Game

In the art of modeling the fight between precision and interpretation and perception of meaning is always there. Simple examples like two-by-two matrices may be combined with verbal discussions to illustrate points in brinksmanship, deterrence, and so forth. But simple analogies can be as dangerous as they are valuable. In chess and poker we all know what constitutes a move; there are well defined rules which tell us what a move is. In diplomacy much of the fun comes in inventing moves which had not been perceived as feasible.

The interpretation of what formal models might mean and the discussion of the perceptual difficulties in modeling is delayed until Section 4; here we concentrate on the exposition of a purely formal model and we put off the various interpretations of the model until Section 4.

Let us imagine a formal game with well defined rules where at the start each player has to make a choice among two alternatives. We call the selection of an alternative a move. Each player's move is decided upon simultaneously. Let us give the players names; we call the first one blue and the second one red. They each commit themselves to their respective moves without knowing what the other has done. After both have committed themselves to their moves they are informed of what has transpired. They are then called upon to each simultaneously select another alternative. Call their original alternatives 1 and 2. If the move of each had been to select his first alternative then in the second stage of this game we assume that they face a choice among two alternatives each. If anything else had been selected as the first set of moves we assume that each player must make his selection among three alternatives. After each has decided upon his second move the game is over and a payoff or reward is given to each player.

The primitive concepts involved in the description above are player, alternatives, choice, moves, information, and payoffs.

We can represent the detailed anatomy of this complete game by a simple diagram called a game tree. Figure 1 illustrates this. The node labeled zero indicates where the game starts. Each node may be regarded as a choice point. Each branch radiating downward from every node (except the terminal nodes) represents an alternative at a particular choice point. Each point is labeled with the name of

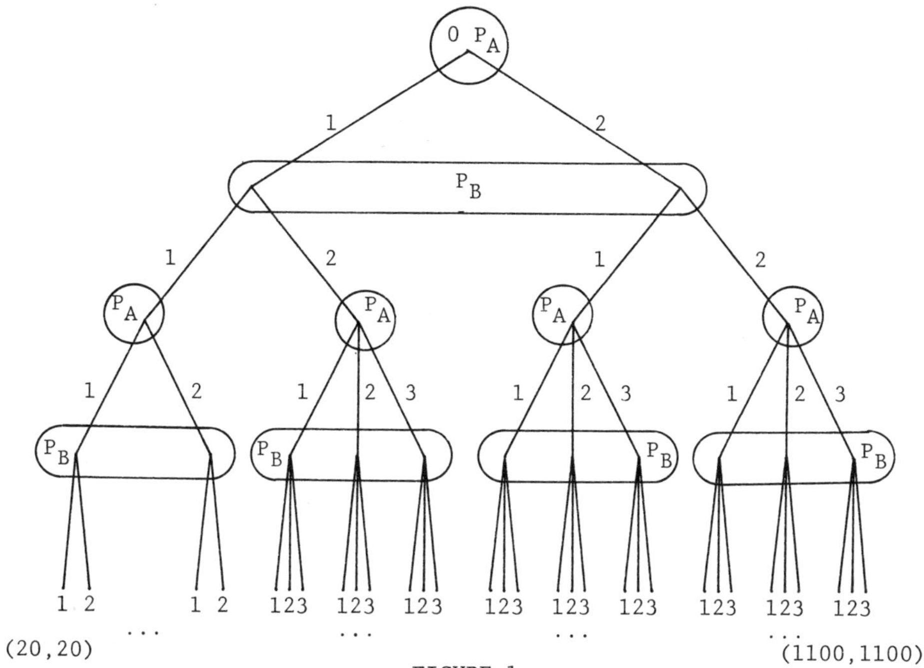

FIGURE 1

the player who is called upon to choose at that position during the game. Thus, here we have ascribed the first choice to player A whose name is indicated by the symbol P_A.

The next two nodes are labeled with P_B and are enclosed by a curve illustrating that they belong to the same information set. An information set (or more properly a "lack of information set") is a set of choice points which cannot be distinguished from each other by the player who is called upon to select his move. Operationally in this setup it does not matter if A or B went first, if neither find out what the other has done before both are committed.

The fact that both are completely informed after each has selected his first move is indicated by the four one-element information sets at the second level (or ply) down the tree. Each of these choice points are labeled with P_A indicating that player A is called upon to select his second move and at that point he knows precisely how the game has developed. In particular, if the selection of moves had been (1,1) then player A is confronted with two alternatives. If the selection of the first moves had been anything else, then player A is confronted with three alternatives. An immediate analog with chess can be seen depending upon the variations in the opening positions at any point in time the number of alternatives among which a player must select his move will vary.

Once more in order to indicate informational simultaneity in the moves we break the information sets of player B into four sets. Each set indicates that player B does not know what was the second move selected by player A but knows all history previous to that point. After B has selected his move the game terminates and both players receive their final payoffs. In Figure 1 if on each occasion each player had selected his first alternative, the final payoffs would have been (20,20). This symbolizes 20 to the first player and 20 to the second player. We have not specified the units. They could be utiles, brownie points, billions of dollars, or many other items. The specification of goals and the measurement of payoffs is clearly a critical problem.

If at each choice the players had initially selected their second alternative and followed that with their third alternative, then the payoff indicated is (-1100,-1100). There should be a pair of numbers at every terminal node. For brevity we have only indicated two payoff values out of the thirty-one needed.

A complete path down a game tree starting at the origin and ending at one of the terminal nodes can be called the play of a game. It is a complete history in full detail of how the action developed from the beginning to the end.

Although it is not done in Figure 1, we can also introduce the possibility of external random events. Even ones where the probabilities will depend upon what the players have done. For example, we can invent a player called P_N who moves at any relevant point during the game with preassigned probabilities. An interpretation of this, for instance, might be that the actions of the two players are to start a theater nuclear war but a random event such as the wind or weather conditions in general might considerably influence the outcome of the first engagement. When probabilities have to be taken into account then the payoffs must be considered in terms of expectations or some evaluation which takes into account the effect of the randomness.

2.3. Strategy and the Normal or Strategic Form

The extensive form outlined in 2.2 displays in detail the sequencing of the moves and the information conditions. For some purposes this detail may not be required. In particular, the game theoretic concept of strategy can be used to remove much of this detail and represent the same situation for different analytical purposes by a payoff matrix stressing only strategies and payoffs. The payoff matrix for the game illustrated in Figure 1 is given in Table 1.

The game theory definition of strategy is not the same as the military definition and the distinction between the two will be explored in Section 4.

In essence, the game theoretic concept of strategy is a complete plan of action covering all contingencies. Another way of looking at it is as follows: imagine that there were a group of agents employed by player A and player B. Suppose that each player has one agent at every information set he has; thus, in the example in Figure 1 player A will have five agents and player B will have five agents. A strategy is a book of instructions which can be given to every agent and which contains information for him as to how to act when called upon to select among his alternatives.

Even in this extremely simple two-stage game there are fifteen strategies or different books of instructions which could be left to the agents.[1]

The extensive form illustrated in Figure 1 gives rise to the 15 × 15 matrix representation of the game in normal or strategic form. The 15 (nonredundant) strategies for player A are as follows:

Strategy 1: "I select alternative 1 for my first move; if (1,1) then I select alternative 1 for my second move. If (1,2) then I select alternative 1 for my second move."

This can be abbreviated as:

"(1,1) then 1; (1,2) then 1." We call this strategy 1. A full listing of the fifteen strategies is given below.

1. (1,1) then 1; (1,2) then 1
2. (1,1) then 1; (1,2) then 2
3. (1,1) then 1; (1,2) then 3
4. (1,1) then 2; (1,2) then 1
5. (1,1) then 2; (1,2) then 2
6. (1,1) then 2; (1,2) then 3
7. (2,1) then 1; (2,2) then 1
8. (2,1) then 1; (2,2) then 2
9. (2,1) then 1; (2,2) then 3
10. (2,1) then 2; (2,2) then 1
11. (2,1) then 2; (2,2) then 2
12. (2,1) then 2; (2,2) then 3
13. (2,1) then 3; (2,2) then 1
14. (2,1) then 3; (2,2) then 2
15. (2,1) then 3; (2,2) then 3

Clearly, if the agents were in fact commanders at a lower level in the hierarchy they would have more degrees of freedom, initiative, and individual action than merely to follow a book of instructions. But at this point our purpose is to make precise one way of conceptualizing strategic analysis. This way clearly will call for modification when considering the influences of perception, aggregation, and organization.

As the game selected here for illustration is symmetric (the false assumption of symmetry is always a danger in modeling--sometimes simplicity comes at too high a price) each side has fifteen strategies. We can number the strategies from one to fifteen where the numbers have no other significance than names. They may be regarded as a listing of the volumes. Volume one contains a certain plan; volume two contains a somewhat different plan and so forth.

Looking at Figure 1 we see that there can be thirty-one different plays of this game. A play is a path from the initial node to a terminal node. There are thirty-one terminal nodes and the path leading to each is unique. There can, however, be 225 combinations of strategies used. Any one of the fifteen strategies of player A can be used against any one of the fifteen of player B. This is shown in Table 1. The numbers in the column at the side are strategies of player A and the numbers across the top are the strategies of player B. The two numbers in each cell represent the payoff to player A, followed by the payoff to player B. For example, suppose that each employs his first strategy. This calls for each to select his first alternative as his first move. But given that the first pair of moves is (1,1) then each will select his first alternative for his

A \ B	1	2	3	4	5	6	7	8	9	10	11	12	13	14	15
1	20, 20														
2		20, 20					0, 0								
3			20, 20												
4				-90, -90											
5					-90, -90										
6						-90, -90									
7		0, 0					-90, -90								
8								-200, -200							
9									-1100, -1100						
10										-90, -90					
11											-200, -200				
12												-1100, -1100			
13													-90, -90		
14														-200, -200	
15															-1100, -1100

second move. Thus, the play of the game will trace a path down to the terminal node on the far left of Figure 1, where we observe the payoffs are 20,20.

How are these payoffs to be calculated for all instances? In order to envision fully the sensitivity of responses we need all 450 numbers in Table 1. Only 34 are indicated, but we show a way for generating all of the others.

Empirically it could be that partial payoffs are obtained at the end of period 1 which are then added to the partial payoffs obtained at the end of period 2 to form the final payoffs. Supposing that this were the case, then a simple way to calculate the payoffs for this table is as follows. If in the first period the moves were:

(1,1) the partial payoffs are (10,10)
(1,2) the partial payoffs are (-100,100)
(2,1) the partial payoffs are (100,-100)
(2,2) the partial payoffs are (-100,-100)

If in the second period the moves are:

(1,1) the partial payoffs are (10,10)
(1,2) the partial payoffs are (-100,100)
(1,3) the partial payoffs are (-1000,1000)
(2,1) the partial payoffs are (100,-100)
(2,2) the partial payoffs are (-100,-100)
(2,3) the partial payoffs are (-1000,900)
(3,1) the partial payoffs are (1000,-1000)
(3,2) the partial payoffs are (900,-1000)
(3,3) the partial payoffs are (-1000,-1000)

Thus the payoffs from strategy 1 used by each are (10,10) + (10,10) or (20,20). The payoffs from strategy 2 used by player A and strategy 7 used by player B are (-100,100) + (100,-100) = (0,0).

It is obvious that for a game with many moves and a fair amount of information the proliferation of potential contingent plans becomes astronomical. The number of strategies for a game such as chess is beyond the bounds of reasonable computation. This does not detract from the precision of the concept of strategy; it merely calls our attention to the fact that when humans attempt to implement a concept such as contingent planning the proliferation of alternatives is so enormous that at best they can only seek a judicious selection of contingent plans based upon perception and aggregation as well as computation.

We have derived the strategic form of the game given in extensive form in Figure 1. When we look at Table 1 we observe that all detail concerning moves and information sets has been removed; whereas merely by the display in Table 1 emphasis is placed on strategies and payoffs.

One can derive the game in strategic form from the game in extensive form because in essence one is throwing away detail. There is no unique way to work backwards. In other words, one cannot uniquely define the extensive form of the game knowing only the strategic form.

<u>Tactics or Strategy?</u> At this point the reader interested in application must be warned about a confusion that could arise. Although it is true that the strategic form of a game can be derived from its

extensive form, as a matter of modeling it may well be that the modeler wishes to take the strategic representation of the game as his primitive concept. This is a matter of modeling choice and in actual application to military problems many tactical situations are easily modeled and analyzed in some variation of the strategic form of a game without ever considering the extensive form. These applications include the study of duels (see Kimeldorf in this volume); the study of games of pursuit and surveillance (see Ho and Oldsder in this volume); and problems in force allocation. A simple example of the last is given below. A mythical Colonel Blotto has three divisions which must be used to defend two forts. An enemy has two divisions to attack the forts. There are four ways in which Blotto can divide his divisions and three ways in which the enemy can divide his. A superior force will capture a fort, but in an engagement of equal size the defenders will win. Suppose that the value of each fort is 100 to the winner and -100 to the loser. The payoff matrix to this simple game is illustrated in Table 2.

	(2,0)	(1,1)	(0,2)
(3,0)	200	0	0
(2,1)	200	200	0
(1,2)	0	200	200
(0,3)	0	0	200

TABLE 2

The four strategies for Colonel Blotto are indicated by the different troop assignments. Thus (3,0) stands for assigning three divisions to the first fort and no divisions to the second fort.

The entries in Table 2 show the payoffs to Colonel Blotto. The payoffs to the other side are the negatives of these amounts as this is a game of pure opposition.

An inspection of the payoff matrix leads immediately to some suggestions. In particular, Colonel Blotto must split his forces and the attacker cannot afford to split his forces. We return to this point when we consider solution concepts in Section 3.

2.4. The Coalitional Form and Negotiation

The cooperative form is utilized when considering the possibilities for negotiation. The concern is focused upon what various groups can claim as their individual "justified demands." In some instances a reasonable way to calculate minimum individual claims is to use a worst case analysis. In this instance referring to Figure 1 and Table 1 neither player A nor B can guarantee more than -1100, whereas together they can obtain 40.

We may symbolize the description of the coalition possibilities by using a set function which assigns a value to each coalition. For n players there will be 2^n-1 values. Thus, for the two players we have:

$$v(A) = -1100 \qquad v(B) = -1100$$
$$v(A,B) = 40$$

When there are only two players almost all of the information that counts is contained in the evaluation of the individual claims and joint benefits. For three or more players the combinations become important. Thus even for five countries in a pact there are already 31 subgroupings which may need to be considered. In Section 3, when solutions are discussed, two three-person games will illustrate how an analysis of the potential gains of coalitions alone can offer some insights into the stability of alliances.

Once more it can be seen that in the progression from the extensive form to the strategic form to the coalitional form of a game, detail is thrown away. There is (as already indicated) a natural way of going from the extensive form to the strategic form and it is unique. There is no "natural way" to go from the strategic form to the coalitional form of a game unless you believe that "worst case" analysis is the correct way to analyze what a coalition can claim. But this may well lead to a virtually paranoid assessment with the implicit or explicit assumption that the rest of the world is out to get you.

In this article no attempt is made to enter into the full details and intricacies of modeling and formal game theory. The reader interested in an extensive overview of models, theory and general applications is referred elsewhere (see Shubik, 1975, 1982; Shapley and Shubik, 1972). Here only a minimal exposition of the basic concepts is given in order to consider them in terms of usefulness and relevance to military and diplomatic, strategic, or tactical analysis.

3. THE SOLUTION CONCEPTS OF GAME THEORY

3.1. Normative or Behavioral Solutions?

The specification of the rules of the game or the complete description of a strategic situation in extensive, strategic, or coalitional form is in and of itself a pre-solution. In other words, an enormous number of assumptions and a considerable amount of analysis has to be made in order to obtain a description of the strategic situation being studied. By the time this has been done already a considerable narrowing of our scope of attention has taken place.

In any form of strategic planning the selection of the model and the specification of its scope is a major task and many of the bureaucratic biases may already be built in at this level. Old diplomats, seasoned generals and politicians all have different perceptual stories to tell in contrast with those of a newly minted Ph.D. in quantitative methods for strategic analysis.

The construction of strategic models is, at best, an artistic and scientific activity heavily constrained by social process. The activities of sponsors, scientists, and advocates are all interwoven.

The construction of strategic models for analysis encompasses the host of empirical work done in index construction, listing of relevant variables, construction of tables of inventories, and preliminary assessment studies.

A reasonable model serves as a data organizing device where perceptions about causality provide guidance for the procurement of information and decisions on level of detail.

The work, however, is not over when one has described the model. If there is any lesson of importance that game theory has to teach, it is the lesson that our usual concepts of individual, rational behavior do not extend to situations involving more than one individual. In other words, even if it were possible to ignore all social psychological and bureaucratic features of existence with safety there still is no nice natural way to extend our concept of individual rationality to situations involving an intermix of community and conflict of interests. Thus, even if we have a full description of "the game" we have to make an inductive leap based upon our social perceptions as to what we wish to consider to be a solution. Full net assessment requires not only that we know what the numbers are, but what the numbers mean. In order to know what the numbers mean we need not only an assessment of capabilities but also an assessment of intentions. The concept of solution broadly speaking involves the deducing of probable behavior from assumptions not only about capabilities but also about intentions. The value of this exercise will clearly be highly dependent upon the quality of the perceptions.

Underlying much of the work in the social sciences and especially the work in economics, operations research, and military analysis has been the development of two parallel views of theory development. One view concentrates on behavioral description of societies and individuals, while the other lays stress upon normative theories. The first tries to offer an explanation as to why people do what they do, while the other offers advice as to what one should do in certain situations.

There is one philosophical school of thought whose proponents feel that there should be a unique normatively most desirable way for individuals and nations to behave. There should be some universally acceptable way of resolving the conflicting goal problem which takes into account both the realities of power and the concepts of equity and efficiency. Eventually this philosopher's stone may be found but in the current state of strategic analysis and of game theory there does not appear to be a unique agreed upon set of assumptions concerning intent or behavior but there are many different solution concepts and a patchwork of partial theories which have been more or less justified in certain usages. In the remainder of this section three solution concepts are noted. Two are in essence normative and the third is usually defended on behavioral grounds.

3.2. Pure Opposition and the Maximin Solution

A two-person zero sum game is a game in which what one side wins the other side loses. It is the very definition of what is meant by opposition. In Table 2 the Colonel Blotto game provides an example of a two-person zero sum game representing a tactical situation.

It was argued by Von Neumann that without having to add too many extra assumptions concerning individual intent or behavior the idea of individual rationality can be extended to a situation with pure opposition and the normative suggestion is that a rational individual should select a strategy which maximizes the expected value of the minimum payoff he will receive by using that strategy. This is simply predicated upon the proposition that an intelligent enemy whose goals are diametrically opposed to the player will be trying to minimize the player's payoff as in that manner he maximizes his own payoff.

In terms of the Colonel Blotto example, the maximin solution for Colonel Blotto involves deciding with equal probabilities to either use his second or his third strategy, i.e. to use two divisions in the first fort and one in the second fort, or vice versa. He must take great care not to signal his intentions to the enemy. The enemy's choice is to concentrate all of his forces and with equal probabilities either hit the first fort or the second fort. He, too, must take great care not to disclose his intentions. If there is an information leak on either side, the game is strategically considerably different.

In the study of dueling and weapons evaluation the maximin solution appears to make good sense. It is a reasonably safe assumption to not count upon your enemy being a fool or more grossly incompetent than you in attempting to evaluate the relative merit of opposing weapons or the chances of detection or evasion in a search game.

In certain situations it may be well known that the morale of one side is low, that their training is poor, or that there are plain fools. But before such an assumption is made it is important to remember that the art of deception is being crazy like a fox.

3.3. Noncooperative Equilibrium

Although it may be reasonable to model certain tactical situations as zero sum games with total opposition, in general most strategic situations and certainly situations involving grand strategy and diplomacy do not have a pure opposition of interests. The game used for illustrative purposes in Figure 1 and Table 1 serves as an example. The fates of the individuals are not related in pure opposition.

What should we consider to be a solution to such a strategic situation? Were we to moralize, we might suggest that in the best of all possible worlds the individuals should jointly maximize, thus in Table 1 the strategy pair (1,1) should be selected or possibly the strategy pairs (2,2) or (3,3).

Tables 3a and 3b provide two simple examples of two-person nonconstant sum games where each individual has just two strategies. If we were to subscribe to the proposition that the parties should jointly maximize in each case they would select their first strategy. But

	1	2
1	5 / 5	-15 / 20
2	20 / 15	0 / 0

a

	1	2
1	5 / 5	-15 / -20
2	-20 / -15	0 / 0

b

TABLE 3

the games are considerably different. In the first game there clearly is a "sucker's bait" incentive for each to move away from the joint maximun. If I know that you are committed to your first strategy then you are a sitting duck for me to move to my second strategy gaining twenty for myself and inflicting a loss of fifteen on you. But the reasoning is circular and if we both do it we may end with zero each. The experimental evidence in laboratories is overwhelming that this is what individuals, in fact, do in one shot experimental games of this variety.

This analysis does not hold for the second game. If both have selected their first strategies then any deviation can only make it worse for both. Here we encounter a paradox. By making certain parts of the environment bad enough for both sides the stability of a jointly desirable outcome may be improved. This is one of the key concepts in discussions of deterrence and in the design of self-policing systems.

Unfortunately, even the simple example shown in Table 3b raises another problem. Suppose for historical or other reasons that both players had been trapped into utilizing their second strategies, unless they can simultaneously decide to switch to their first strategies an independent move could cause damage to both. The moral of this example is that there are instances in which the self-policing of the system becomes so good that it conceivably could lock the system in to an undesirable as well as a desirable state.

Table 4 shows yet another type of situation. Here strategies (1,1) and strategies (2,2) are both favorable. The first pair is more favorable to the first player and the second pair more favorable to the second player. The problem is clearly one both of coordination and conflict.

	1	2
1	20,10	0,0
2	0,0	10,20

TABLE 4

The concept of a noncooperative equilibrium point has been in economic literature since the writings of Cournot in 1836. In a more general form it has been in game theoretic and strategic analysis since the work of Nash in the 1950's.

A noncooperative equilibrium point has the property that it is self-policing against any individual violation. Thus, for example, in the game illustrated in Table 3a if I know that you are committed to strategy 2 it is my self interest to play strategy 2 and vice versa. Thus the strategy pair (2,2) resulting in the payoffs of zero to each is in equilibrium.

This solution concept offers some plausible explanation of the stability of some outcomes but it does not provide any dynamics nor does it offer unique predictions.

It does not have the normative property that outcomes will necessarily be optimal to all or even any of the participants. Perhaps the most attractive aspect of the noncooperative equilibrium solution theory is that it does not presuppose cooperative behavior and that in certain situations there seems to be some evidence that it offers a better explanation than other solutions concepts. But possibly even more important the noncooperative strategic analysis of relatively small games provides analogies to strategic and bargaining situations illustrative of many problems in strategic analysis involving escalation, deterrence, preemptive strikes, precommittment, doomsday machines, the deliberate introduction of the probability of failure or uncertainty into certain systems, and other items.

Modeling and mathematical attempts to extend the noncooperative equilibrium analysis to multistage problems have resulted in a closer appreciation of the close relationship between any sort of dynamics and behavioral and organizational problems. One immediately has to take into account not merely information, memory, and learning but also signaling, communicating, teaching, and deceiving.

A detailed discussion of the various aspects and modifications of noncooperative equilibrium theory is given elsewhere (see Shubik, 1982, Chs. 8, 9 and 10).

3.4. Cooperative Solutions

Cooperative solutions are primarily based upon the coalitional form of the game and frequently use the characteristic function to provide the initial description of individual and joint claims.

There are many competing solution theories at this time; most of them normative, but some behavioral. Among the better known solution concepts are the core, the value, the bargaining set, and the stable set. The first two are usually given normative interpretations, whereas the last two have been offered as explanations of the outcomes of bargaining behavior or of social stability.

A full description of these competing solution concepts together with examples is given elsewhere (see Shubik, 1982, Chs. 7, 8, and 11). In this exposition only the core is discussed and illustrated with two simple three-person examples. Even at this level of simplicity these examples and this solution concept serves to indicate a fundamental problem in alliance stability and a paradox concerning the gains to be had from cooperation and the sum of individual claims to these gains.

Tables 5 and 6 show two three-person games in coalitional or characteristic function form. The interpretation of Table 5 is given in detail, the interpretation for Table 6 is similar.

$$v(1) = v(2) = v(3) = 0$$

$$v(12) = 1,\ v(13) = 2,\ v(23) = 3$$

$$v(123) = 4$$

TABLE 5

$$v(1) = v(2) = v(3) = 0$$

$$v(12) = 2,\ v(13) = 3,\ v(23) = 3\tfrac{1}{2}$$

$$v(123) = 4$$

TABLE 6

The three players are called 1, 2, and 3. The first three values of the characteristic function indicate that an individual acting alone can obtain zero (the zero should be interpreted not necessarily as nothing but as his status quo point or even the payoff from a worse case analysis). If players 1 and 2 can form a coalition without 3, they will obtain 1. If 1 and 3 form a coalition without 2, they obtain 2. If 2 and 3 form a coalition without 1, they obtain 3. If all players together decide to cooperate then jointly 1, 2, and 3 can obtain 4.

Given cooperative behavior then the three players will have to decide as to how to split the prize of 4 between them. The core solution to a game suggests that the players should cooperate in order to maximize joint gain and then they should split the gain in such a way that the claims of all subgroups are satisfied. In other words, no subgroup gets less than it could have obtained by breaking the alliance.

For an n-person game we call a set of n non-negative numbers an <u>imputation</u> if it adds up to the total gain obtainable by all n players. An outcome of a bargaining agreement will be in the core if it is an imputation and gives all groups as much or more than their "go it alone" claims. For the general three person game an imputation will be in the core if it satisfies the following set of conditions:

$$\alpha_1 + \alpha_2 + \alpha_3 = v(123)$$

$$\alpha_i \geq v(i) \text{ for } i = 1, 2, 3$$

$$\alpha_1 + \alpha_2 \geq v(12)$$

$$\alpha_1 + \alpha_3 \geq v(13)$$

$$\alpha_2 + \alpha_3 \geq v(23)$$

Checking these conditions for the game shown in Table 5 we can see that the imputation (2/3, 4/3, 6/3) lies in the core. Altogether they obtain 4; individuals all get more than zero; 1 and 2 obtain 2; 1 and 3 obtain 2-2/3; and 2 and 3 obtain 3-1/3. It is relatively straightforward to check that there are many other agreements or ways of cutting the cake which are in the core.

A quick check of Table 6 shows that the suggested division of (2/3, 4/3, 6/3) is not in the core. In fact it is possible to show that there is absolutely no way whatsoever in which the alliance of all three can split their joint gains in such a way that every pair will obtain at least what it claims to be able to get by going it alone.

The existence of a non-empty core tells the analyst that there are possibilities for agreement where all subgroup claims can be met. When the core is empty this cannot be done and the motivation for defection of groups who perceive that they can do better by going it alone is high.

Much of the value in utilizing games in coalitional form comes in the estimation of the values for the characteristic function. That exercise alone requires a careful assessment of the different joint interests of members of an alliance. For example, Germany, Holland, and Belgium or Norway and Denmark might have group assessments which should be specially recognized when evaluating an alliance including all of them, others, and the United States.

3.5. External Symmetry and Strategic Analysis

> The natural qualities of a warlike people play just this part: bravery, aptitude, powers of endurance, and enthusiasm.
>
> —Clausewitz, p. 256

War is a complicated matter and in the analysis of war or for that matter any human process it is extremely easy to leave out of consideration critical variables. Morale, bravery, national will, terrain, weather, and a host of other factors all may appear to be important.

The constant fight in the construction of operational models is between confusing oneself with irrelevant detail which adds to the appearances of reality but subtracts both from analysis and understanding, and the leaving out of critical factors.

In strategic analysis and in weapons evaluation or doctrine development there are some natural divisions of the variables and other factors which can and should be made. In particular most human factors are taken as given and unless otherwise stated explicitly the factors are assumed to be the same for all. Thus, for example, in describing a tank fight between a South and North Korean tank unless otherwise stated the implicit assumption is that the competence, morale, and intelligence of the tank crews are the same.

In all of the formal modeling of game theory the assumption of <u>external symmetry</u> is present as a key modeling convention. It can be phrased as follows: unless explicitly noted otherwise all unspecified factors are assumed to be the same for all sides. If we do not explicitly introduce differences in fighting ability or morale; if we fail to explicitly note that equipment failure and repair characteristics are noticeably different, then our analysis will proceed on the basis that the differences do not matter.

If the differences do matter, then we should blame ourselves not for bad analysis but for bad initial perceptions and poor modeling.

Methodology is not a substitute for perception. If used with care, it may, however, provide a supplement to perception. If one knows how to model fast and analyze clearly, an analysis which reaches ridiculous conclusions may serve as a warning that one's initial perceptions were poor and this in turn can lead to a re-conceptualization of the problem at hand.

4. MODELING, THE STRATEGIC AUDIT, AND ANALYSIS

4.1. Systems Modeling and Modeling Systems

The art and science of systems modeling, analysis, and net assessment take place within a bureaucratic context. The mere fact that we are dealing with operations removes the process from the sole domain of the research institutions and academies.

The bureaucratic environment provides much of the motivation or volition and many of the biases to perception which go into the model building process.

Although this article is not directly concerned with the study of the organizational aspects of strategic decision making, three references are suggested for those who wish to consider in more detail the social processes influencing strategic planning. The references are the article "Optimal Ways to Confuse Ourselves" by Albert Wohlstetter (1975), The Adviser by Herbert Goldhamer (1978), and Victims of Group Think by Irving Janis (1972).

Concerning the debate on strategic arms Wohlstetter offers advice on how to save a dogma, how to keep from noticing that your predictions have been wrong, how to ignore the decline in strategic spending, and other advice as to how to bend perception during debate.

The lengthy scholarly essay of Goldhamer deals with the role of advisers in political systems, the forms in which advice is given, the background of the advisers, the risks involved in being an adviser, and the relationship between advice giving and decision making.

The work of Janis helps to characterize the basic way committees and groups constituted for decision making are able to bias their joint decisions. He suggests that among the more important aspects to be guarded against are 1) a sense of invulnerability; 2) shared stereotypes; 3) agreed upon rationalizations; 4) a comforting group shared illusion of morality; 5) self-censorship; 6) an illusion of unanimity in decision making; and 7) a level of mind guarding that guarantees that off-beat ideas will not be considered.

In a highly amusing and instructive article entitled "A Guide for the Perplexed," Amrom Katz (Air Force/Space Digest, November 1967) offers eight maxims to help guide tourists through a classified bureaucracy. The first four maxims appear to be particularly relevant both to the science and process of strategic planning.

Maxim 1: where are the calculations that go with the claculated risk?

Maxim 2: inventing is easy for staff outfits. Stating a problem is much harder. Instead of stating problems, people like to pass out half accurate statements together with half available solutions which they can't finish and which they want you to finish.

Maxim 3: every organization is self-perpetuating. Don't ever ask an outfit to justify itself, or you'll be covered with fact, figures and fancy. The criterion should rather be, "what will happen if the outfit stops doing what it's doing?" The value of an organization is easier determined this way.

Maxim 4: try to find out who's doing the work, not who's writing about it, controlling it or summarizing it.

4.2. On Modeling

In the remaining sections we concentrate on the art and science of modeling applied to strategic problems rather than to the political process.

There are at least five aspects of model building which must be considered at any level of analysis. They are: 1) abstraction and analogy; 2) omission; 3) aggregation; 4) symmetry; and 5) simplification and approximation.

Clausewitz offers an analogy between war and economic competition. Recent work in political science has been devoted to analogies between party contests and economic competition. How fruitful and how general are the analogies? When confronting a new situation it is highly desirable to look around to see if a precedent has already been established. Was there a piece of history which comes close to where we are now? Is there a different subject where the phenomenon has been studied under a different name? Will a re-labeling of the major variables indicate a kinship in causal structure to a different problem?

Perhaps one of the hardest tasks in modeling is to be able to have a conscious justification for at least a few major factors left out of current consideration. For example, we are only going to consider the Democrats versus the Republicans even though there is a third party candidate running because traditionally third party candidates never matter. Or we only need to consider the posture of the Soviet bloc against the Western bloc because it is ridiculous to conceive that a third party trying to provoke a war would sink the Nimitz. Or previous general staffs were foolish to have ignored the tank and aircraft as interesting toys but of little military significance, but Laser technology is not like those examples it really is in the realms of science fiction and can be safely ignored for at least several decades.

Although high speed digital computers can manipulate many variables, human beings and organizations use extensive aggregation in their conceptualization of problems. Thus, the behavior of quarter of a billion individuals is summarized into the behavior of a macro player called the Soviet Union. The aggregate noun "they" is invariably used to describe the battle between the individual and the surrounding bureaucracy. The selection of the appropriate level of aggregation involves both the choosing of the appropriate macro players and the aggregation of economic, political, social, and other factors. Clausewitz, for example, tended to more or less ignore or to aggregate the logistical features of war.

Throughout all of the social sciences, throughout philosophy, and mathematics the assumption of symmetry is beguilingly attractive. Problems which have many forms of intrinsic symmetry are frequently easier to analyze than those which do not. There are, unfortunately two major dangers with assumptions concerning symmetry. First, the assumptions are frequently wrong and are made from either an implicit wish to simplify without empirical justification. The second problem is somewhat more subtle. Even if the problem is intrinsically symmetric it is possible that there are solutions which are not symmetric. Yet, one's mind set will tend to be such that non-symmetric solutions to symmetric problems may be overlooked.

Simplification and approximation are called for in all operational analysis. Especially where our insights are shaky and the structure of the environment is both complex and uncertain. Linear, quadratic and other simple approximations will tend to abound. Simple indices will be constructed. Not because we have undying faith in their portrayal of the world but because we should be able to justify that they are better than nothing and are about all we can handle.

4.3. A Simple Escalation Model

In Figure 1 a relatively simple example of an abstract game in extensive form was presented. There were two players, two periods, and depending upon what took place during the first period there were two or three alternatives during the second period. The game description and notation was purposely kept completely abstract to serve as an example of a precise notation. A key act in modeling is to map from real problems onto abstract structures to be analyzed. A concrete interpretation of this model is offered together with a critique of some of the difficulties encountered in modeling.

This can be regarded as an extremely simple stripped down version of a Soviet bloc-Western bloc confrontation with danger for escalation to theater and possibly strategic warfare. We may assume that player A is the Soviet bloc and player B is the Western bloc. This assumption involves an enormous aggregation and simplification which may or may not be reasonable depending upon the question to be answered.

The two alternatives faced by each player during the first period can be interpreted as 1) maintain the status quo peacefully, and 2) attempt to extend one's influence by means of a theater war. The implicit assumption or omission here is that there would be no inclination for either side to go directly to a strategic war without going through the intermediate stage of having a theater war start.

During period two if no war has started the alternatives remain peace or theater warfare. If a theater war has already been instigated then the alternatives are peace, the continuation of the theater war, or an escalation to strategic war. In this model it is assumed that the game ends at this point. In reality extreme care has to be exercised in specifying terminal conditions. The perceptive work of Ikle (1971) on Every War Must End is of critical reference at this point.

The information conditions assumed are such that each side is expected to be able to maintain information security concerning its intentions at about the same level.

This simple model is laced with symmetry assumptions. It is probable that this may be the point of greatest lack of realism. Is it pious thinking or empirical fact that leads us to assume that the Eastern and Western bloc players have equal capabilities and equal views and values on the possibility of war?

Another critical difficulty that can be seen in the inspection of this model is that in the second round a clear perceptual distinction has been made between the choice of theater war and strategic war. It may well be that one of our gravest difficulties is to know at what point a theater war has become strategic.

In Section 2.3 a method for calculating the payoffs resulting from the employment of any pair of strategies was given. It was done by calculating the period payoffs for all states of the system during period one and then again during period two. In making the analogy between this game and an escalation situation we have to ask is it even a reasonable first approximation to hang numbers onto these payoffs, let alone to add them? Where do these numbers come from, who is responsible for generating them, what empirical inputs can be used, and what moral decisions must be made when the calculations implicitly or explicitly involve the evaluation of mega-deaths or other forms of mass destruction?

When the players are gross aggregates it is extremely hard to attach simple single numbers to their goals. There are many other more complex preference structures which can be considered (see Shubik, 1982,

Ch. 5). But the problem here is not purely scientific. It is precisely here where the moral aspects of war call for a social as well as scientific assessment of how high a price is being paid.

Applying the concept of the noncooperative equilibrium solution to this model one such equilibrium has both sides risking a theater nuclear war which then is escalated into a strategic nuclear war. The model is limited and the strategic reasoning is somewhat simplistic but if only as a parable it has a story to tell which needs to be considered. In this model as formulated the guards required for an adequate self-policing system are not there. Is it merely a fault of simplistic modeling or is it a fact of life?

The argument here is that the very exercise of formal modeling if done with emphasis more on meaning, concepts, and causality provides an important tool for improving the perceptions and making more precise the questions.

The model, its analysis, and its answers are not an end in and of themselves. They are part of a process leading to better questions, better understanding, and communication between analysts and those with operational responsibility which in turn leads to more sophisticated and better models and analysis.

4.4. The Strategic Audit

An elementary consideration of the problems involved in modeling human affairs is enough to suggest that the peculiarly rationalistic and relatively precise concepts of strategic analysis and game theory must be modified and interpreted with great care when modeling actual tactical, strategical, or grand strategical situations. It is suggested here that the conceptual organization provided by the languages of game theory combined with the less precise but more historical and socially oriented language of military strategy and diplomacy enable us to construct a checklist which serves the purposes of structuring a strategic audit. This provides a structure and a checklist for observing and relating the key factors which must be taken into account when constructing models for strategic analysis.

The major headings for such a checklist are: 1) scope; 2) time frame; 3) players or strategic actors; 4) rules of the game and choice sets; 5) payoffs and goals; and 6) behavioral assumptions.

These headings, in terms of the discussion presented in Section 2 provide the modeling outline for translating one's understanding of the nature and scope of the problem at hand in to the formal rules of the game.

A more detailed breakdown of these six distinctions is given.

1. Scope
    a. Grand strategic; military diplomatic
    b. Theater strategic
    c. Tactical
2. Time Frame
    a. Date of study
    b. Length of period studied
    c. Event oriented, fixed clock or both
    d. Initial conditions assumed
    e. Terminal conditions assumed

3. Players or Strategic Actors
   a. Level of aggregation
   b. Number of strategic players
   c. External symmetry assumptions
   d. Verification of "strategic dummies"
4. Rules of the Game and Choice Sets
   a. Relevant political detail (own and others)
   b. Relevant bureaucratic detail (own and others)
   c. Relevant technological knowledge (own and others)
   d. Information and communication conditions
   e. Scope of feasible actions
5. Payoffs and Goals
   a. Short and long term
   b. Defined on a finite or indefinite horizon
   c. Ordered, partially ordered or other
   d. Quantifiable (risk measures)
   e. Team or individual goals
   f. Principal agent or fiduciary goals
   g. Zero sum or nonconstant sum environments
6. Behavioral Assumptions
   a. Rationality
   b. Risk behavior under time pressure and/or stress
   c. Instinctive or programmed behavior
   d. Problems of perception and interpretation

Under scope the three broad distinctions made by Jomini are adopted. They are, in essence, grand strategic where military, diplomatic, and national goals and actions are all intermixed. Strategic and theater strategic where national and diplomatic goals can be held constant while our consideration turns to overall military planning; and tactical where we are limited to a specific direct engagement of less than global scope.

Although the need to specify the date of the study and the length of the period to be studied is fairly obvious, nevertheless it needs to be stressed that assumptions concerning what items can be ignored or treated as parameters depend delicately upon the length of time selected and to some extent the specific date. For example, seasonal factors may be important.

There is no magic overall rule which distinguishes the short term from the long term. But an <u>ad hoc</u> decision must be made for any model concerning the influence of time.

Another important distinction is between event oriented and fixed clock oriented models. The game tree does not have a time metric imposed upon it in general, but illustrates the sequence of decisions. Many routines are oriented towards clock time whereas many important decisions are oriented towards specific changes in the environment which can take place at any point and which must serve as cues for action.

Virtually all planning models cut out a slice of time from an ongoing process. A true understanding of the dynamics implicit in even the most limited one shot model requires careful specification of both initial and terminal conditions. The initial conditions postulated in and of themselves summarize the modeler's view about the relevance of history.

The terminal conditions postulated recognize the possibility that there may be a tomorrow.

A technical trick often used by mathematicians to produce paradoxical results is the backward induction. One starts at the end of time when there is only one choice left for each side, figures out what the best action should be under those circumstances and works backwards to figure out what you should do now. This device has some merit in forcing explicit thought in long range planning about the future but it depends delicately upon being able to specify plausible terminal conditions which model the fact that there will probably be a future history which is being reflected in the assumptions made about the terminal period.

One of the fundamental weaknesses in the simple escalation model presented in Section 4.3 is the assumption that there are only two time periods. Whenever there is a tomorrow there is a chance that you may be punished for what happened today. In military conduct the hoards of Genghis Khan appeared to be highly conscious of the possibilities of tomorrow. They tended to kill traitors and informers on the basis that if they could be false to one master now they could be false to them later. They also tended to massacre deserting armies on the basis that a desertion at one time could be followed by a desertion at another time.

The institutional difficulties in extricating oneself from an adventure commenced without contemplating its alternative ends are enormous. The difficulties in coping with how to end wars have already been noted. In economic matters it is probably true that getting out of bad businesses or bad investments is harder than getting in to good ones.

Much of strategic analysis whether military, economic, or political has been based upon models of purposive behavior. In order to stretch the domain of definition of purposive behavior most of us anthropomorphize organizational behavior. Thus we talk about the French morale is low, England expects, the Soviet Union demands, and so forth. As the size of the aggregation grows the problems with defining goals and purpose of the aggregate player increase considerably.

Our attention span is relatively limited. It is easier to think about matters if we can aggregate the players highly and limit the number of players. Frequently the most attractive limit is two. "Us and they." Most tactical analysis is done with two players and is cast into a zero sum or confrontation mode. It is tempting but almost always false to do the same with strategic analysis. This becomes especially dangerous when studying the grey zone between strategic nuclear warfare and theater warfare. In spite of its horrendous implications much of the first and second strike analysis has been carried out at virtually a tactical level mode of thought. Yet, the odds are overwhelming that the commencement of a strategic war would be preceded by a theater war. In this case in the selection of relevant players extreme care must be exercised in removing even Albania or Argentina from the list.

In Section 3.5 sufficient discussion has been given concerning external symmetry conditions. No further comment is made here.

The verification of strategic dummies is a technical problem frequently confronted in game theoretic analysis which is of importance in strategic analysis in general. In situations involving more than two parties it is exceedingly difficult to pre-judge when a particular third party will have no influence of significance on the outcome. It may be safer to include the third party into the analysis even at the cost of more work and then by the analysis be able to demonstrate that the third party is irrelevant and can be left out of further analysis with safety.

Depending upon the level of analysis political, bureaucratic, and technological detail will be more or less important. Especially when planning at the level of grand strategy the assumptions concerning these factors will tend to be made by different groups and depending upon their professional biases the analysis will tend to be biased accordingly.

The old adage from Bridge that "a peek is worth two finesses" holds true for both war and business. In Section 2 it has been indicated that the language of game theory as used in describing the extensive form is particularly adapted to displaying information conditions. It must be stressed, however, that the meaning of information both in game theory and information theory is devoid of semantic content. The statement that you have more information is equivalent to the statement that you can perceive more states of the system not that you understand what these states mean.

How to model both information and communication conditions in the description of strategic situations is still one of the most difficult tasks which face the strategic analyst. Many of the skills of bargaining and negotiation depend upon making distinctions fuzzy and dissolving perceptions of opposition. The more tactical the situation is the easier it becomes to characterize information. Whether a certain bridge has been blown up or not can be described with reasonable precision.

The description of the scope of feasible actions is in terms of formal game theory the equivalent of specifying the strategy sets for all players. At best the description must be highly aggregate. The astronomical proliferation of strategies in the formal game theoretic sense, i.e. totally encompassing contingency plans, make it quite clear that the choice of alternative plans to analyze must be highly limited.

Those who use numbers and mathematical analysis in strategic planning tend to like one number descriptions of payoffs. In all of the examples given in Section 2 the formal game theoretic models had one number representing the payoff to each player or in the case of the games in coalitional form one number was used to represent the size of a joint gain which a group of allies could divide between them.

There are many different technical ways in which complex goal structures may be described when presenting a strategic analysis. The techniques are there, but the difficulties are primarily conceptual and empirical. What is a good representation of the goals of an organization? How does risk behavior where the individual is risking his own life or assets differ from risk behavior where he may be sending others to their death or ruin? No formal methodology is going to answer these questions. On the other hand it can and does assist in asking them.

The last set of conditions we must consider in modeling are our behavioral assumptions. A complete description of the strategic environment does not tell us how our competitors or opponent will behave. Whether we like it or not the process of strategic analysis involves implicitly or explicitly the piling of assumption upon assumption. At this point most strategic studies make some form of assumption about individual rationality and the analytical abilities of the other sides. Unfortunately this is not enough to be able to determine behavior. In Section 3 above it has been noted that there are many solution concepts to nonconstant sum games of strategy. Phrasing this less technically it amounts to observing that there is no unique generalization from the commonly accepted concepts of individual rationality to multi-person rational behavior. The paradoxes encountered in some game theoretic models tell us that tunnel vision individual rational behavior by all parties may in fact spell joint disaster.

In general the very act of strategic planning is purposive. The overall context in which it takes place is behavioral. Throughout strategic analysis there is an element of goal oriented normative planning buried in a sea of instinctive or programmed behavior. The good planner has to make do with assumptions that are at best more or less precise. He needs to make behavioral assumptions about large segments of his own system and the organizations of others but he cannot afford the luxury of ignoring the element of purposivity and calculation that lies in his own decisions and the decisions of others.

4.5. Tactics, Strategy, and Negotiation

> Tactics is the theory of the use of military forces in combat. Strategy is the theory of the use of combats for the object of the war.
>
> In a word it is easier to make a theory for tactics than for strategy.
>
> —Clausewitz, pp. 173, 190

> Diplomacy is to do and say
> the nastiest thing in the nicest way.
>
> —Isaac Goldberg, The Reflex

The three quotations above are supplied as a preamble to some closing remarks on modeling at the three different levels of tactics, strategy, and negotiation.

The formal methods of game theory do not distinguish between strategy and tactics. This distinction is made at the level of modeling.

In essence the limitations of scope, and time; the specification of the two players and their goals all make the two-person zero sum formal game model of tactical warfare a reasonably good representation of the problem at hand. Be it situations involving dueling, weapons evaluation, search, or force structure there are many reasonably tight models which can be formulated. The excellent article by Weiss (1953) on "Requirements for a Theory of Combat" presents an exposition which applied equally well to Lanchester equation, game theory, and other models of combat.

With the advent of nuclear warfare and intercontinental ballistic missiles a new category of analysis has appeared. For want of a better description following a distinction made by Jomini at the level of strategy we distinguish grand tactics from tactics. It is suggested that first and second strike analysis or the "salvoe study of nuclear war" can best be regarded as tactical analysis on the grand scale. The difference between conventional tactical analysis and this is that the game is not zero sum, the payoff functions are vague in the extreme, and the terminal conditions are difficult to conceptualize let alone describe. A highly simple form of this analysis is given by Dalkey (1965) in his paper "Soluble Nuclear War Models." The modeling of the command and control system for the first twenty minutes after a verified red alert and the subsequent twenty-four hours is an activity that hardly exists in the open literature. Yet its implications and importance are enormous.

As we proceed from strategic through to grand strategic analysis the time span increases, the perception of the key variables become vaguer, the moral imponderables become greater, and the setting of values and goals becomes more and more a part of a social process than an empirical given of the problem.

As diplomatic considerations begin to intermix with the more purely military aspects of the situation, the need for accurate inventories of weapons and what they can do still remains but the conceptual framework must include bargaining and negotiation as well as force. The analysis of coalitional structures is valuable in indicating as a first approximation what there is to bargain about. At this point three psychically different approaches can be distinguished. The first is that the study of bargaining and diplomacy should be left to non-quantitative students of political science, diplomacy, and history. The second is exemplified by a growing literature on formal models of bargaining processes. A brief volume by Roth (1979) provides a coverage of axiomatic models of bargaining.

The third approach presents an intermix of history, political science, game theory, social psychology, law, and mathematical analysis. This is typified by the work of Ikle (1964) on How Nations Negotiate; Schelling's (1959) Strategy of Conflict; and Fisher's (1970) little handbook on How to Negotiate.

In all of these the mixture between the analytical and behavioral approaches can be seen. Frequently tied in with the rational actor view of strategic negotiation is the normative proposition that principled problem solving should be able to resolve even international strategic conflict.

Perhaps there may be a rational analytical way for solving international conflict, but if there is it will depend upon the development of international law which in turn depends upon the expansion of our mutual understanding of the strategic choices we face.

FOOTNOTE

[1] In full detail there are $2\ 2\ 3^3 = 108$ strategies including redundancies. However in a system prone to errors all 108 may be relevant.

REFERENCES

[1] Clausewitz, K. von, On War (Pelican Books, London, 1832, 1968).

[2] Dalkey, N., Soluble nuclear war models, Management Science, 11 (1965) 783-791.

[3] Fisher, R., How to Negotiate (Harper & Row, New York, 1970).

[4] Fisher, R., International Conflict for Beginners (Harper & Row, New York, 1970).

[5] Goldhamer, H., The Adviser (Elsevier, New York, 1978).

[6] Ikle, F. C., Every War Must End (Columbia University Press, New York, 1971).

[7] Ikle, F. C., How Nations Negotiate (Harper & Row, New York, 1964).

[8] Janis, I. L., Victims of Group Think (Houghton Mifflin, Boston, 1972).

[9] Katz, Amrom, A guide for the perplexed, Air Force/Space Digest (November 1967).

[10] Roth, A. E., Axiomatic Models of Bargaining (Springer Verlag, New York, 1979).

[11] Schelling, T., Strategy of Conflict (Harvard University Press, Cambridge, 1959).

[12] Shapley, L. and M. Shubik, The rules of the game, in: Game Theory in Economics, Ch. 3, Rand Corporation Paper R-904/3-NSF.

[13] Shubik, M., Games for Society Business and War (Elsevier, Amsterdam, 1975).

[14] Shubik, M., Game Theory in the Social Sciences (MIT Press, Cambridge, 1982).

[15] Weiss, H. K., Requirements for a theory of combat, Ballistic Research Laboratories, Aberdeen Proving Grounds MR 667, 1953.

[16] Wohlstetter, Albert, Optimal ways to confuse ourselves, Foreign Policy, 20 (1975) 170-198.

MATHEMATICS OF CONFLICT, Martin Shubik (ed.)
Elsevier Science Publishers B.V. (North-Holland), 1983

A CRITICAL ASSESSMENT OF QUANTITATIVE METHODOLOGY AS A POLICY ANALYSIS TOOL

Ralph E. Strauch

The Rand Corporation
Santa Monica, California

I. INTRODUCTION

Recent decades have seen a tremendous expansion in the role and influence of quantitative methods and techniques in the governmental decisionmaking process. Operations research and related techniques found fertile fields of application during World War II, and their wartime successes were sufficient to guarantee them a continuing place in peacetime defense planning. Robert McNamara's tenure as Secretary of Defense, with its emphasis on systems analysis, program budgeting, cost-benefit analysis, etc., served further to institutionalize the use of quantitative methods, models, and techniques in defense planning and in the presentation of the results of that planning to Congress and the public. Concurrently, the use of quantitative methods and techniques spread to other departments of government, and to state and local governments as well. These methods increasingly pervade all areas of governmental planning and operations

That use of these methods has, on occasion, proved fruitful and rewarding is without question. That it has, on occasion, proved misleading and led to unfortunate choices and consequences is also without question. The question of the net effect of quantitative methodology on the process of government is a complex one, probably not susceptible to definitive resolution. Some supporters of quantitative methodology and its application see it as the wave of the future --the natural extension of rationality and the scientific method to the process of government. Others, including this writer, are somewhat less sanguine.

This report provides a critical assessment of the usefulness of quantitative methodology as a policy analysis tool, and sketches the outlines of an approach to analysis and the application of quantitative methodology that might ameliorate, somewhat, the major weakness highlighted by the assessment.

That weakness (discussed in Section II), stems from the fact that the theory from which the methodology is drawn assumes the existence of a well-defined problem, and assumes that the model used in analysis is an adequate and accurate representation of that problem. Based on that assumption, it prescribes methods for the logical analysis and solution of the problem defined by that model. Human judgment, per se, is not represented in or treated by the methodological theory.

Many of the problems encountered in policy analysis, however, are highly "squishy" and ill-defined. They may be equally well represented (though poorly, at best) by a number of distinct models. In such situations, the logical analysis of the model may be of relatively minor importance in determining the conclusions reached compared to the highly subjective activities of formulation--choosing a particular model from among the many possibilities available--and interpretation--deciding what the results mean in the light of the model chosen and the deficiencies present in it.

The absence of explicit treatment of these activities in the methodological theory often leads to their neglect in applications of the methodology. This neglect shows itself in a tendency to focus on calculation and the results of calculation and to ignore the issues of what is being calculated, and why. When this occurs, methodology becomes a poor and inadequate substitute for judgment, rather than a useful adjunct to it.

The approach to analysis that might ameliorate the problem (discussed elsewhere) involves open and explicit acknowledgment of the role of subjective human judgment in any analysis of "squishy" and ill-defined problems. The use of mathematical models in analysis is a special case of the more general cognitive function of using models (or various types) to understand more complex phenomena. This report presents a discussion of that function and the beginning outlines of a theory of judgmental analysis.

II. POLICY ANALYSIS AND QUANTITATIVE METHODOLOGY

This assessment is not directed at quantitative methods and techniques *per se*. These are only tools, and one can no more fully understand the application of quantitative methodology in policy analysis by looking at the tools employed than one can fully understand the nature of gardening by examining rakes, hoes, and shovels. The subject of this assessment is the application of quantitative techniques and methods, with emphasis on the manner in which the tools are applied--the attitudes, beliefs, etc., of analysts applying the tools, and the approaches to problem solving and to the interpretations of results that those attitudes and beliefs engender. These issues of belief, approach, and interpretation are subtler, and far more difficult to deal with, than are the superficially straightforward and direct issues of method itself. They are, nonetheless, equally important in determining the value of quantitative methodology as a tool for policy analysis, and the worth of the knowledge and advice produced using it.

Policy Analysis

As used in this assessment, the term "policy analysis" should be broadly construed to connote the systematic examination or analysis of questions arising from, related to, faced by, etc., the governmental planning and decision making process, conducted with the intention of affecting or contributing to that process. It thus includes forms of analysis such as force structure studies, program evaluations, etc., that might be excluded by a narrower use of the term. It includes studies done within the Air Staff, the Department of Defense, or other portions of the government, as well as by outside study agencies such as Rand when those studies are intended to affect governmental policy and operations

The problem encountered in policy analysis run the gamut from well-defined to highly "squishy." A *well-defined* problem, in this sense, is one that can be given a clearcut, well-defined formulation, amenable to rigorous analysis. "How big a rocket is needed to put a man on the moon?" for example, is such a problem. A "*squishy*" problem, on the other hand, is one with the property that any clear cut, well-defined formulation of it will look like an unambiguous representation of the substantive problem only so long as we don't lean too hard on it, or question it too carefully or deeply. Thus, for example, the question "What strategic forces are needed to deter nuclear war with the Soviet Union?" is a very squishy question. The "Assured Destruction" formulation of that question, which measures the adequacy of a candidate strategic force in terms of its ability to inflict damage on the Soviet Union in a spasm exchange, is one way of turning it into an apparently well-defined problem. The appearance is only superficial, however, and evaporates rapidly when probed to any great extent. As a general rule, the well-defined problems encountered in policy analysis tend to be those concerned almost exclusively with technological or physical questions. Problems with any significant degree of behavioral or political content tend to be squishy. The more central the behavioral or political content becomes to the substantive issue, the squishier the problem is likely to be.

Policy analysis is carried on for a number of varied and sometimes conflicting objectives and purposes. Among these are the choice of a particular course of action to implement a predetermined policy, the advocacy or support of a particular position or point of view in the process of intra-governmental debate, and the search for illumination or insight into the nature of a particular issue or problem in order to better inform the policy- and decisionmakers who must deal with that problem. This last objective--of providing illumination and insight--is, in many ways, the most difficult and demanding. This is because in squishy problems there is no clearcut way of deciding how much illumination is enough, or measuring how much insight a particular analysis really provides. The present assessment is concerned primarily with this objective and with problems at the squishy end of the problem spectrum. For this reason, some of the criticisms made here may appear to be overstatements when related to better defined problems and less ambitious objectives.

The social sciences deal with many questions similar to those arising in policy analysis, but I will not be concerned with social science methodology here. The distinction between the two, as far as I am concerned, is one of the objectives for which knowledge is sought. Social science reflects an academic interest in knowledge, while policy analysis reflects a more operational interest in supporting the ongoing functions of government. We may, therefore, impose problem-independent validity criteria on what we consider social science knowledge and simply recognize questions for which answers meeting those criteria cannot be found as being beyond the purview of scientific knowledge. Policy analysis, however, must deal with questions as they occur, and must provide understanding on the basis of existing information. Its criteria cannot be problem-independent since it does not have the choice of rejecting problems for which the criteria cannot be met.

Quantitative Methodology and Its Supporting Theory

As used here, the term "quantitative methodology" refers to the existing body of mathematical methods, computational techniques, etc., and supporting theory available to the policy analyst. This includes the theory of and tools derived from probability theory, game theory, statistical inference, econometrics, decision theory, computer simulation techniques, operations research, systems analysis, etc. These are overlapping terms with no clear dividing line between them. Each, in some sense, contains the others.

This methodology has an abstract existence and meaning--as a body of mathematical knowledge about ways of dealing with theoretical problems defined in and examined by the supporting theory--that is independent of, though related to, its application to substantive problems. The utility of the methodology as a tool for application to substantive problems depends on the existence of appropriate similarities between the theoretical and the substantive problems. In order to assess the application of the methodology to policy analysis, it will be useful to begin with a brief examination of the theory, *per se*, and of the nature of the problems it addresses and the solutions it provides to those problems.

The supporting theory can be classed as mathematics, *per se*, in the sense that it is a study of mathematical systems and models as abstractly defined objects in their own right, independent of their relationships to real systems or real world substantive problems. The structure and behavior of mathematical model as such is determined fully by the defining premises and in no way by "real" considerations about any substantive process being modeled. Description of abstract models in "real world" terms, however, often serves as a useful heuristic device for understanding their behavior.[1]

Probability theory, for example, is the study of abstract mathematical structures called "probability spaces" or "measure spaces." Methodologies that derive from probability theory, such as statistical inference or decision theory, arise from the study of particular classes of problems that live on, i.e., are defined on or in terms of, those probability spaces. The problem of estimating the regression coefficients in a multivariate Gaussian process is a well-defined and meaningful abstract problem regardless of whether anything approximating a multivariate Gaussian process exists in the real world. Conversely, the fact that there exist "optimal" methods for estimating regression coefficients in such a process confers no necessary validity on those methods as tools for dealing with real problems involving data produced by something other than a multivariate Gaussian process.

Mathematical analysis, at the theoretical level, consists of the exploration of the structure and behavior of mathematical systems and models as logical consequences of the defining premises of those models. The results thus produced are grounded in the defining premises and connected to them by a chain of logical inference. The product of such analysis is depicted schematically in Figure 1. It consists of a mathematical model, the results derived from the model, and the chain of logic connecting the results to the model.

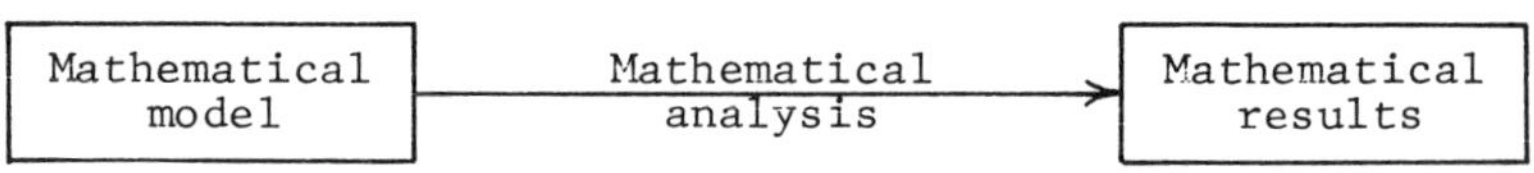

FIGURE 1. Mathematical Analysis

The model is a well-defined mathematical object whose structure and behavior is logically determined by its defining premises, and, perhaps, by additional particularizing data (parameter values, boundary values, input data, etc.) that differentiate the particular instance or case under consideration from the broader class defined by the model. For example, a statistical model might describe the behavior of a statistical sampling process, with particularizing data describing the observed sample in a particular case. The results are precise statements about the further behavior or structure of the model, or about the theoretical problem defined by the model, that follow logically from its hypothesized structure. The grounds for the results consist of a chain of logical inference connecting the premises defining the model with the results derived from it. This chain may include (manual or computer) arithmetical computations since these are a form of logical inference.

The results produced by mathematical analysis are "objective" in the sense that they are grounded in a set of well-defined premises and logical inference from those premises. The structure shown in Figure 1, if it is fully articulated, provides a representation of the results and the grounds on which they rest that might be said to be "free-standing," in the sense that their validity can be determined on the basis of that structure alone, with no reference to or dependence on the competence or judgment of the analyst producing it. As such, they are subject to independent verification or repudiation by an outside critic on logical grounds alone. As we will see later, this is a property not shared by many of the applications of quantitative methodology arising in policy analysis.

It sometimes appears from the free-standing, objective nature of mathematical results and their grounds that mathematical analysis is a purely logical activity, with little or no intuitive content. This is not the case. While the validity criteria of mathematics require the existence of a logical chain of proof that can be verified with no reference to intuition, initial discovery of that chain of proof is frequently a highly intuitive process. Good mathematicians develop a high degree of intuition about the mathematical models and theories with which they deal. They use this intuition in analyzing mathematical models and deriving results from them, particularly in producing original and creative results. The logical validity checks that mathematics imposes on those results, however, effectively "factor out" the role of intuition in the final product--producing the objective, free-standing product described earlier. This fact notwithstanding, creative mathematical analysis is a highly judgmental and intuitive process.

Rigorously Quantifiable Problems

The application of quantitative methodology to a substantive problem involves the use of a mathematical model as a simplified representation of the problem, or of some phenomenon important to the problem.[2] The substantive problem or phenomenon is analyzed through the analysis of the model. The nature of the conclusion reached, and the amount of credence and confidence that can be placed in it, depends in part on the results produced by mathematical analysis of the model. It also depends significantly on the relationship between the problem and the model--on what parts of the problem the model represents and how well, and on what parts of the problem the model distorts or fails to represent and how badly. Strictly speaking, this latter aspect of application--the effect of problem/model relationship--falls outside the scope of the theory on which the methodology is based since the theory deals with analysis of the problems defined by the models treated in the theory. This matters little when the problem/model fit is a good one, but can become important when the problem differs significantly from the model used to analyze it.

The simplest applications, from the viewpoint of the problem/model relationships involved, are those in which the structure and logic of the substantive problem are the same as (at least relative to the question being addressed) the structure and logic of the model being used. Such problems will be referred to here as "rigorously quantifiable." Rigorously quantifiable problems arise, for example, in the analysis of physical systems that obey well-understood and accepted scientific laws or in statistical experimentation in which the analyst may introduce randomness into his experiment to match the randomness in his statistical model. Analysis of rigorously quantifiable problems is methodologically pleasing because of the ease with which the theory transfers to application. The coincidence of structure between problem and model insures that the theoretically founded statement about the model will have valid analogs as statements about the problem.

Analysis of a rigorously quantifiable problem is depicted in Figure 2. The analysis has two levels, a substantive level and a mathematical level. The analyst begins with the substantive problems on the substantive level. He adopts a mathematical model of that problem and analyzes the model to produce mathematical results. The model may not be spelled out in detail, particularly if it is commonly used and accepted for the type of problem he is dealing with (Euclidean geometry, for example, as a model in surveying or map making). It is understood to be there, nonetheless, and in principle, to be capable of rigorous specification. The results are then projected back up to the substantive level as substantive conclusions.

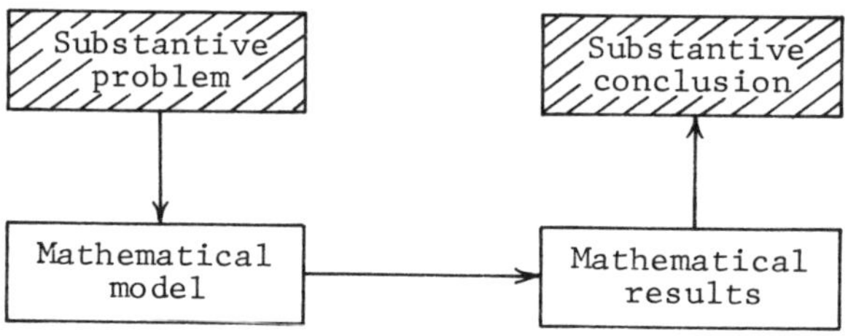

FIGURE 2. Analysis of a Rigorously Quantifiable Problem

This is not to imply that the process of analysis is necessarily the straightforward, orderly process depicted here. Rather, when the process is completed, the end product can be sufficiently neatened up that it will appear as though it had been.

The calculation of the trajectory of an artillery shell is an example of a rigorously quantifiable problem. Newtonian mechanics is a mathematical model of "real mechanics" that has a well-defined structure independent of any relationship to reality. It seems to fit reality well (at least in nonrelativistic situations) and is commonly used as a surrogate for reality in analysis. To find the trajectory of an artillery shell used the Newtonian model of the problem, the analyst plugs in the particularizing data--weight of the projectile, force of the charge, elevation of the gun, wind, etc.--and solves the trajectory problem within the model. The analyst then accepts the result obtained as a substantive conclusion about the real trajectory of the real shell. This can be done, in fact, without ever consciously thinking about the fact that Newtonian mechanics is not "real mechanics" at all, but simple a mathematical abstraction that serves as a convenient surrogate.

A statistical experiment using random sampling is another example of a rigorously quantifiable problem. In this case, the analyst assures himself of a good model/problem agreement by structuring the problem to fit the model. He does this by designing a sampling procedure (the problem) in a way that insures that it possesses the random character required by the statistical model that will be used to analyze the sample. The analyst can thus accept the model as a surrogate for the actual sampling procedure and accept the results derived from the model as conclusions about the substantive problem.

The validity of conclusions produced in this way depends on two factors: (1) the internal logical validity of the mathematical analysis and (2) the empirical validity of the linkages between model and substantive problem. The criteria for logical validity are the same as in the case of theoretical analysis discussed earlier. The difference between mathematical analysis and application lies in the fact that the premises defining the behavior of the model are no longer simply logical premises, but are assumptions about the behavior of the real world in the substantive problem. As such, they must be empirically validated. In problems dealing with physical systems obeying well-understood and accepted scientific laws, the validation is provided by the fact that the models used represent those scientific laws. In problems of statistical experimentation, empirical validity may be achieved by making the problem look like the model. In other cases, it may simply depend on the existence of long empirical experience with the process, sufficient to justify the treatment of the model as a surrogate for the process. This might be true, for example, of queuing models used to study the behavior of telephone exchanges.

If these logical and empirical validity criteria are met, then the conclusions reached are "objective" in the sense of being grounded in rigorous logic and objective fact. The structure of conclusions and grounds shown in Figure 2 is, like that of Figure 1, free-standing in the sense that its validity can be decided on the basis of the structure as it stands, without reference to or dependence on the competence or judgment of the analyst producing it. (This is true, at least, if the structure is fully articulated. It is true only in principle otherwise.) It is, thus, subject to verification or refutation by an independent critic on the grounds of the logic of the mathematical analysis within the model and the empirical connection between the model and the substantive problem.

Also, as in the case of the mathematical analysis, analysis of rigorously quantifiable problems conveys the false impression of being a strictly logical, rather than intuitive, activity. As in mathematics *per se*, intuition and good judgment on the part of the analyst may play a major role. But also as in mathematics, they are filtered out by the free-standing structure of verification and proof. Thus, while judgment may play a significant role in the formulation of the problem and in the conclusions reached, that judgment plays no part in the ultimate grounds for the validity of those conclusions.

Other Applications

Rigorously quantifiable problems possess a structure sufficiently like that of the models used to represent them to justify the straightforward interpretation of results derived from the models as conclusions about the problems. If all problems to which quantitative methodology were applied were like this, there would be no need for concern about the limits of the methodology, or about applications not clearly justified by theory. That is not the case. Most of the problems to which quantitative methodology is applied in defense planning, and indeed in the analysis of public policy more generally, are not rigorously quantifiable problems. Rather, they span a spectrum from what might be called <u>reasonably quantifiable</u> problems at one end to highly <u>squishy</u> problems of dubious quantifiability at the other. Significant policy problems, in particular, frequently tend to lie much nearer the squishy end.

By a <u>reasonably quantifiable</u> problem, I mean one that possesses a fair degree of natural structure that seems to be reasonably represented by the mathematical model used to analyze the problem. Examples might include logistics problems such as inventory or maintenance scheduling problems. In many ways, analysis of reasonably quantifiable problems looks very much like the analysis of rigorously quantifiable problems. The major difference is that the requirements for empirical validity of the link between problem and model are less stringent in the case of reasonably quantifiable problems. The main criterion is that of reasonableness not of scientific validity.

A highly <u>squishy</u> problem, on the other hand, is unlikely to have any clear-cut, well-defined formulation that is both analytically tractable mathematically and that unambiguously captures the substantive problem. Clear verbal statements of the substantive problem may appear analytically intractable, while analytically tractable formulations of the problem are likely to significantly distort, or at least restrict, its substance. As a general rule, problems with significant behavioral or political content tend to be squishy. The more central that behavioral or political content becomes to the substantive issue, the squishier the problem is likely to be.

I want to look critically at this spectrum of problems and at the usefulness and the limitations of quantitative methodology across this spectrum. My emphasis will fall more heavily on the limitations of the methodology and on the squishy end of the problem spectrum--a combination often neglected in the advocacy literature for quantitative methodology. For this reason, some of what I have to say may appear overly critical when taken to apply to the application of quantitative methodology to reasonably quantifiable problems.

The extension of Figure 2 to the application of quantitative methodology more generally is shown in Figure 3. As in Figure 2, there is a substantive level at the top, consisting of the substantive problem and the conclusion the analysis provides about that problem, and a mathematical level at the bottom, consisting of a mathematical model and the results obtained from the model using mathematical analysis. Unlike the rigorously quantifiable case, however, the model is rarely a direct match for the problem nor the conclusion a straightforward translation of the results. As the figure attempts to illustrate, both the substantive problem and conclusion may be far from sharply defined. The model itself will still be sharply defined, at least in principle, since that is the nature of mathematical models. This is true even if the model is not completely specified but is left partially implicit, since the mathematical results have rigorous meaning only with respect to a well-defined model. Suppose, for example, that a statistical test is applied and that the results are stated in terms of the "statistical significance" of the data. The use of a statistical model within which that term has meaning is implied, even if it is not explicitly stated.

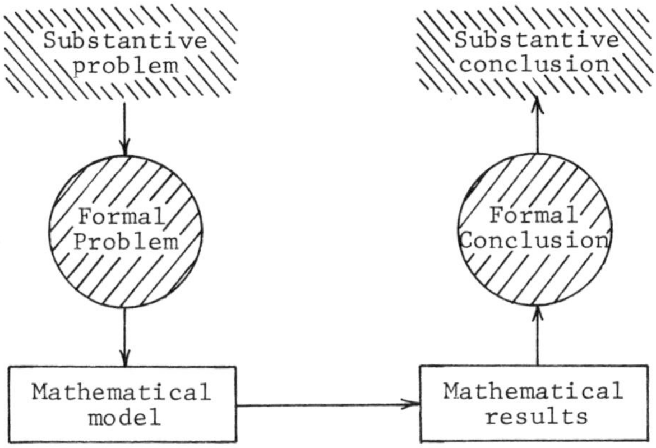

FIGURE 3. Application of Quantitative Methodology:
the General Case

The relationship between the substantive problem and the model may be highly tenuous and ambiguous--particularly if the substantive problem itself is very squishy. The links between problem and model, and corresponding links between results and conclusion, may be far from clear. For this reason, it is useful to think of an intermediate formal level between the substantive and mathematical levels, serving to link the two. The formal problem links the substantive problem and the model by specifying what is being modeled in substantive terms. It delineates the parts of the problem addressed by the model, the assumptions being made, the parts left out, etc. It delineates the relationship of the model to the problem, perhaps describing or restating the model in substantive terms. The formal conclusion serves similarly as a link between the results and substantive conclusions--interpreting the results in a context more closely tied to the substantive problem and less to the model.

The formal problem and formal conclusion may not be explicitly identified and separated from either the substantive problem or model. They are, nonetheless, implicit in any application of quantitative methodology to a less than rigorously quantifiable problem and play an important role in determining the relevance and validity of the overall analysis. As Figure 3 attempts to show, the structure of the formal level will typically be more sharply defined than that of the substantive level, but may still be considerably less precisely defined than the mathematical level.

If the substantive problem is reasonably quantifiable, as in the case, say, of an inventory stockage problem, then the formal and substantive problems may merge together and assume a structure almost identical with that of the model. As we move along the spectrum in the direction of increasing squishiness, however, the three become more clearly distinguishable, if not always more clearly distinguished.

Consider, for example, the comparison of alternative candidates for ground support aircraft to meet U.S. commitments to NATO, using a computer simulation of a NATO conflict. The substantive problem is one of determining which aircraft would better serve U.S. needs in the NATO theater. The model is the mathematical abstraction of a NATO/Warsaw Pact conflict embedded in the computer program used for the simulation. The formal problem in this case might be taken to be the evaluation of the aircraft in the specific conflict scenario of which that mathematical model is a model. The results produced by the computer simulation provide a measure of how well each of the abstractions of aircraft embodied in the model perform in the abstraction of war embodied in the model. These results can then be translated into a formal conclusion about how well the real aircraft would perform in the real conflict represented by the model. This translation of results into formal conclusion embodies judgments about the relationship between the model and the formal problem. Translating this formal conclusion into a substantive conclusion about the relative value of each of the aircraft involves additional judgments about the relevance of the particular conflict scenario considered in the formal problem to U.S. security interests in NATO.

Analysis can be thought of as consisting of the following three components:

1. Formulation: Formulation includes the choice of the formal problem and of the model to be used to represent that problem --the activity along the downward links on the left side of Figure 3.
2. Mathematical Analysis: This occurs at the mathematical level, at the bottom of Figure 3. It consists of analysis of and within the context defined by the model. This analysis produces results as mathematical statements about the model. It may also include purely technical questions of formulation and interpretation arising within the model itself, as opposed to those involving the relationship of the model with the formal and substantive problem.
3. Interpretation: This includes interpretation of the results as a formal conclusion, and of that conclusion as a substantive conclusion--the activity along the up links on the right side of Figure 3.

In practice, these components are seldom separate, sharply defined activities occurring in sequence. Rather, they blur together as interwoven aspects of a complex and intellectual process. If they become separated, it is only when the results of that process are neatened up at the end. The formal problem and conclusion may not be explicitly identified as distinct from the model and the results or from the substantive problem and conclusion. Indeed, there is frequently no need for separate explicit identification, particularly in the case of reasonably quantifiable problems. Whether they are articulated as such or not, however, a formal problem and conclusion are tacitly assumed in the way results are presented and interpreted.

The validity of the conclusions reached depends on the logical validity of the mathematical analysis and on the validity of the linkages between problem and model. As is true with the cases considered earlier, the logical validity of the mathematical analysis can be determined objectively, without reference to the subjective judgment of the analyst. This is not true, however, of the validity of the linkages between problem and model. In general, no objective standards exist by which these linkages can be validated. The thing that separates the rigorously quantifiable problems from those that are not, wherever they lie on the spectrum of squishiness, is the fact that the role of judgment along the links of formulation and interpretation can be factored out in the former but not in the latter problems. The conclusions reached at both the formal and substantive levels thus remain grounded in the subjective judgment of the analyst.

Formulation and interpretation are essentially subjective activities --requiring and depending on careful and considered judgment on the part of the analyst doing them. Formulation, from the substantive to the formal problem and from the formal problem to the model, is a process of taking away--of removing pieces to make the problem smaller and more analytically tractable. With complex and squishy problems particularly, this requires an intuitive understanding of the substantive problem as well as of the methodology being used. It may also involve some adding on, in the form of assumptions that are questionable on substantive grounds, but that make analysis easier. The assumption of statistical independence of various types of events, for example, falls in this category, as do assumptions about the "rationality" of political decisionmakers. Interpretation, conversely, involves putting things back--adding in the considerations removed to make the problem tractable and removing any distortions resulting from the added simplifying assumption.

The potential for distortion can be seen by considering Assured Destruction (AD) strategic force posture analysis--a form of analysis that has played a major role in U.S. strategic force planning over the last decade. The substantive problem is one of evaluating the deterrent capability of alternative strategic offensive forces. The mathematical model is a "worst case" strategic force exchange calculation in which Soviet strategic forces are applied to do maximum damage to U.S. strategic offensive forces, and surviving U.S. forces are then applied to Soviet value targets, i.e., cities. The results produced are estimates of Soviet fatalities. If these fatalities exceed a specified AD level, the force posture under consideration is deemed to pass the AD test. If not, it is not.

One of the assumptions made in moving from substantive problem to model is the choice of the spasm war embodied in the model as the form of strategic conflict in which strategic forces will be tested. Does this represent a considered judgment that spasm war is the only strategic conflict worthy of substantive interest, or is it a simplifying assumption to make the problem more analytically tractable? This question is of more moment than it might initially appear since it has a major impact on the interpretation of the AD test as a force planning tool.

Should AD be viewed as one measure of adequacy or as the design goal for strategic forces? In other words, should AD be the predominant or only criterion by which we evaluate our strategic force posture, or should it be simply one measure of "How much is enough?" with additional evaluation and comparison of other force characteristics also relevant to any choice between alternatives? The AD criterion came into being in a period (the early 1960's) when the United States had an overwhelming strategic superiority. This superiority resided, moreover, in a force consisting of weapon systems developed and designed for different functions. It can be argued that the initial role of the AD criterion was more to say "How much is enough?" than to specify the design and composition of the strategic forces. Over time, however, it seems to have increasingly taken on the role of primary design goal for strategic force design, at least in the minds of many analysts. This is not the place to argue the substantive merits of this issue one way or the other, but simply to point out that it is, on the one hand, a critical substantive issue of U.S.

defense policy, while on the other, an issue that can be decided by default in the formulation and interpretation of strategic force analyses.

A related but distinct question is that of what substantive assumptions about Soviet perception and behavior are reasonable. The simplest interpretation of the AD calculation in deterrence terms assumes a Soviet leadership constantly motivated to go to war with the United States and deterred only as a result of explicit calculations of war outcomes. This leadership might be characterized as being advised by a "gnome in the basement" of the Kremlin who runs an AD calculation every Friday night and will advise attack on Saturday if expected Soviet fatalities fall below the AD level. The formal problem associated with this formulation is that of deterring that gnome. The AD calculation then has a direct interpretation as a formal conclusion --he will or he won't be deterred. This conclusion, however, requires careful additional interpretation in any broader context in which more realistic Soviet motivations and behavior are considered, e.g., in SALT considerations.

Alternatively, the formal problem can be viewed as one of ensuring that the outcome of total and unconstrained strategic warfare will appear unacceptable to any reasonable Soviet leader. With this formulation, the AD calculation is an ill-fitting but nonetheless useful model. The results produced by the calculation then require more careful interpretation as a formal conclusion, but at the same time, provide more insight into the larger substantive problem.

Formulation and interpretation are inherently judgmental in nature, and the conclusions reached are inextricably grounded in them. Hence, the conclusions themselves are judgmentally grounded. The degree of judgmental grounding, and its criticality to the validity of the conclusions, will vary with the problem. It will generally be greater for squishier problems. The fact of this judgmental grounding is inescapable, and should give lie to the claim that quantitative methods applied to squishy problems produce objective conclusions. The models and methods are tools, and their utility in complex problems depend on the skill and judgment of the analyst using them. To expect, as some do, that the methods themselves necessarily product valid results is no more reasonable than to expect that a particular paint brush will produce a beautiful painting, or a particular knife a fine carving.

The theory from which the methodology is drawn is concerned with the application of the methodology to the problems treated by the theory --problems specified as well-defined mathematical models with which the theory deals. Questions of how the methodology might be applied to problems outside that class (as most squishy problems are) are themselves beyond the scope of the theory and not considered by it. Some theories may address broader ranges of problems than others, but the inherent limitation remains. Many of the squishy problems arising in defense planning and analysis are far more complex than those dealt with in the theoretical constructs. The methodological literature tends to focus on example problems to which the methodology is applicable, with occasional examples of problems to which it clearly is not. The applicability (or usefulness) of the methodology in the gray areas in between (where most interesting applications fall) are treated largely with hand waving and reference to the exercise of good judgment by competent analysts and decisionmakers. Very little guidance is given, however, in the exercise of that judgment. The questions of formulation and interpretation are themselves complex, squishy, and ill-defined questions. It is, perhaps, easier and neater (from a methodological perspective) to pretend that they simply don't arise or that, when they do, they will somehow take care of themselves.

Another factor that contributes to the tendency to ignore the role of judgment in the application of quantitative methods is the fact that this role does not appear important in the area of inquiry in which the application of the methodology has proven most successful--the natural sciences. The problems addressed by the natural sciences are seen as rigorously quantifiable problems. The application of quantitative methodology in policy analysis is viewed by many of its proponents as an emulation of scientific principles and the scientific method. The fact that the methodology provides objective knowledge in science is seen as evidence that it should provide "objective" knowledge in other fields as well. This premise appears largely grounded in faith, however, and seems more a perversion of than an emulation of the principles of science.

This is not to imply that quantitative methodology is of no value in dealing with squishy problems. On the contrary, many examples exist in which the analysis of squishy problems using quantitative methodology has provided significant insights or valuable solutions. This is usually due more to the skill and insight of the analyst, however, than to the power and rigor of the methodology _per se_. It stems not from the fact that he made calculations, but that he saw the _right_ set of calculations to make, and was able to _interpret_ them creatively. Nonetheless, the conventional dogma surrounding the application of quantitative methodology tends to attribute success to methodology _per se_, rather than to the wisdom and judgment of the analyst using it. It suggests that appropriate standards by which to judge quantitative methodology should focus on the technical quality of the mathematical analysis, rather than on the quality of the judgment associated with the formulation and interpretation of the analysis. These standards, in turn, encourage considerable amounts of sloppy analysis and confound the problem of separating the good from the bad.

Method-Oriented Analysis

One result of inadequate attention to the role of judgment in the application of quantitative methodology is what might be called method-oriented analysis. In effect, the analyst pretends to be in the situation represented in Figure 2, rather than that shown in Figure 3. A model is chosen that is analytically tractable, and that can be tied to the substantive problem in a plausible way. The methodology is then applied in a straightforward manner, as if the model were a good fit to the problem and no judgmental issues were present in formulation or interpretation. The results are then given a straightforward interpretation in substantive form. The question of problem/model fit is finessed by the observation that conclusions follow from assumptions, and it is up to the readers, customers, etc., to decide whether or not they wish to accept the assumptions. The analyst is saying in effect, "Here's the problem I solved and my conclusion about it. Take it or leave it."

One major failing of the method-oriented approach is its almost exclusive emphasis on similarity between problems and models. It presumes, in effect, that if enough similarities can be found, then the methodology is applicable and will yield good results. This premise, however, is true only if there do not exist enough differences to invalidate it. Unfortunately, for the method-oriented analysts and their customers, however, the existence of many similarities does not preclude the existence of significant differences, as the following example illustrates.

<u>Example 1</u>: <u>Regression Analysis</u>. Consider a set of observations on two time series, X and Y, as shown in the scatter diagram in Figure 4. Suppose that X is a policy variable, in the sense that it measures something subject to policy control, and that Y is a variable that measures something we would like to control. Suppose still further that common sense suggests that the relationship between the two is real, and not specious. What can be said on the basis of the data about the relationship between the variables and about the effect on Y of policy manipulation of X ?

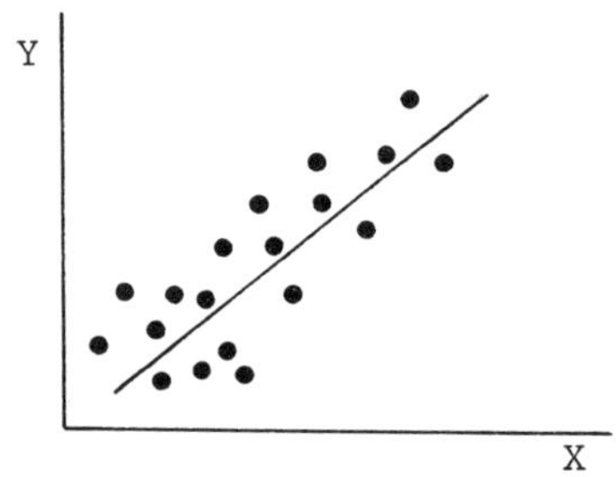

FIGURE 4. (X,Y) Scatter Diagram

For the method-oriented analyst whose method is regression analysis, the answer is clear. Fit a regression line to the data, as shown in Figure 4, and use that line as a predictor of Y given X . Interpret the slope of the line as a measure of the incremental effect of changes in X on Y , and the correlation between observed and predicted values of Y (i.e., the corresponding points on the regression line) as a measure of the expected accuracy of the prediction thus obtained. Using the method-oriented approach, then, it is possible to estimate the relationship between X and Y and to evaluate the effect on Y of policy manipulation of X on the basis of the available (X,Y) data--without worrying particularly about the judgmental questions of what X and Y really measure, what mechanisms really relate them, etc.

But how reasonable are the conclusions thus obtained? The answer to that question requires more knowledge about the problem than was specified above, or than a method-oriented analyst usually takes into account in addressing it. Suppose, for example, that X is the mean annual depth of a reservoir, and that Y is the annual rainfall in the area. These variables have the characteristics outlined above. Because reservoir management policies can be changed, reservoir depth is a policy variable subject to policy manipulation. Annual rainfall is a variable we might be interested in controlling, and common sense clearly suggests that the relationship between rainfall and reservoir depth is nonspecious. If reservoir management policies have remained constant over time, then we might well see a strong correlation between rainfall and reservoir depth.

In spite of this, however, the conclusion suggested by the analysis

--that we can decrease rainfall by draining water more rapidly from the reservoir and maintaining it at a lower level, and can increase rainfall by allowing more water to accumulate in the reservoir and maintaining it at a higher level--seems ludicrous. This is because the causal relationship assumed in the analysis--reservoir depth causes rainfall--runs counter to our common sense understanding of the relationship between the two--rainfall determines reservoir depth.

Because the nature of the mechanism connecting rainfall and reservoir depth is so self-evident, the speciousness of the conclusion is also self-evident. When there are more variables involved, the relationships between them less well understood, etc., the same kind of common sense test of validity may not exist. The plausibility of method-oriented analysis, coupled with a lack of any clear cut basis for disbelief, may then appear as strong grounds for acceptance of the proffered conclusions. A number of analyses of the Viet Nam conflict were conducted in the late 1960's that rested largely on arguments closely parallel to these. This example, however, shows how specious such conclusions can be in the absence of careful evaluation of the fit between problem and model and interpretation of results in light of that fit, i.e., of questions of substantive formulation and interpretation shown in Figure 3.

An equation of the type found in regression analysis, say, such as,

$$Y = aX + b + e \,, \tag{1}$$

where a and b are constants and e is an error or disturbance term, can be assigned two distinct types of meanings. The first is a statistical meaning--the equation describes a statistical relationship between X and Y, viewed as the joint output of some otherwise unspecified process. The statistical meaning refers only to the joint distribution of the variables, and not to the causal mechanism relating them. The equation can also be given a causal meaning in the sense that it describes a system in which the output Y is produced by inputs X and e. There is no necessary connection between the two meanings. To see this, note that the statistical meaning of the equation is unchanged if the equation is transformed into

$$X = a'Y + b' + e' \,, \tag{2}$$

where a', b', e' are appropriate linear combinations of a, b, and e. The causal meaning, however, is reversed.

Relative to the mathematical systems that such equations are conventionally understood to describe--multivariate Gaussian processes--there is no need for distinction between the statistical and causal meanings. Statistically, the equation describes the joint distribution of random variables X and Y. Causally, an (X,Y) pair having this distribution can be thought of as being obtained by sampling X from its marginal distribution, then sampling Y from its conditional distribution given the observed value of X. This mathematical system is equally adaptable to the alternative causal interpretation provided by equation (2)--simply reverse the order of sampling.

When equations such as (1) are used as models of natural systems, however, a distinction between these two meanings is necessary. If X is reservoir depth and Y is rainfall, then equation (1) is a meaningful description of a statistical relationship between them, but not of a causal relationship. Statistical techniques such as tests of significance apply to the statistical meanings of equations such as equation (1). Hypotheses arising in substantive policy problems, however, are more often causal than statistical in nature. An equation such as equation (1) (or its more complex multivariate extensions) can be used to describe a natural system, and a statistical test can be applied to a causal hypothesis about that system. The validity of that test depends on the validity of the identification of the causal and statistical meanings of the equation relative to that natural system. That identification requires separate distinct judgment made outside the context of the test itself. If that judgment is not made, the risk of conclusions equivalent to "reservoir depth causes rainfall" is high.[3]

This type of error can result from confusion in the meaning of words, as well as equations. To say that data are "statistically significant" says only that the data do not appear to have been produced by the mathematical system specified as the model for the "null hypothesis." Leave aside for the moment questions of how closely that system parallels the state of the natural system described by the null hypothesis--e.g., questions such as the randomness of sampling or the linearity of relationships. Even apart from such questions, there is no a priori reason to equate "statistically significant" with "operationally significant" without careful ad hoc judgment relative to the question being addressed. Yet this distinction is frequently blurred.

Another oft confused term is "independent variable." The following are among the meanings it seems to have:

1. On the right of the equal sign.
2. A cause in a cause-effect pair.
3. Chosen first in the sequential selection of a multivariate sample point (the process described above).
4. A variable over which we have some policy control, e.g., we can control reservoir depth by changing our reservoir management policies.

These meanings are all distinct. They may or may not coincide in a particular situation. Assuming a greater degree of coincidence than is reasonable between them is a major cause of fallacious statistical inference. Particularly pernicious is the oft made (usually implicitly) assumption that 1 and 4 are sufficient to guarantee 2. It is this assumption, for example, that leads to the inference that reservoir depth causes rainfall and to other less obvious but equally specious inferences in more complex situations.

Quantitative analysis is largely a process of manipulating symbols--numbers, equations, mathematical statements, etc. If the equations, statements, etc., have different meanings in different parts of the process, then the validity of the manipulations made will be critically dependent on the coincidence of those meanings in the particular problem under analysis. If they fail to coincide, the conclusions may rest on the same illogic and confusion of meanings that, when applied to the word "cardinal," would support the conclusion that the St. Louis ball team belongs to the hierarchy of the Roman Catholic church. The symbols involved may include large amounts of multivariant numerical data, rather than a few simple natural language sentences, and the manipulations consist of detailed computer analysis using complex multiple regression programs rather than simple verbal syllogism. The same underlying fallacy can occur at either level.

In neither case, moreover, is the issue of possible confusion of meanings one that can be settled within the context of the formal analysis--the symbols being manipulated and the manipulations being made. It can be settled only through separate distinct judgment made apart from that manipulation.

Quantificationism

In spite of its limitations, there is a contemporary school of thought that advocates the method-oriented application of quantitative methodology to a broad range of behavioral and political problems arising in policy analysis. It seems to hold that the issues of formulation and interpretation are either nonexistent or will somehow be automatically taken care of by the competence and good judgment of analysts and their customers, even when never explicitly surfaced and considered. This school of thought, which will be referred to here as quantificationism, seems to hold that quantification is a positive value per se, i.e., that for most questions, a quantitative answer is a priori better than a qualitative one. Its more extreme proponents seem to hold the view that any problem of substance is reducible to a form that admits a quantitative solution and that no problem or issue can be adequately understood until it is reduced to such a form. Its more moderate adherents might reject this extreme position. They would nonetheless argue that quantification almost always improves understanding and decisionmaking, and that as a result, quantitative analyses should be conducted whenever possible.[4]

Quantificationism comes in a variety of forms, but its salient features are an emphasis on methodology per se, i.e., on models and techniques and the results that they produce, and the belief that this emphasis is appropriate in the application of quantitative methodology to squishy problems--that it captures the scientific method and provides a scientific approach to such problems. It appears to do this because it appears to emulate the reductionism inherent in the physical sciences.[5] It can be argued, however, that this reductionism should not be considered as the central core of the approach to knowing inherent in the physical sciences, but rather as a characteristic of that approach as applied to the particular problems that the physical sciences address. In attempting to learn from science about how to understand other classes of problems, therefore, we must question whether or not that particular characteristic of science is one that should transfer to this new class of problems.

Rationales for quantificationism are of two basic types--literal and pragmatic. The literal rationale derives from the assumption that the methodology does in reality what it claims to do in theory. The pragmatic rationales, on the other hand, acknowledge the limitations of the literal rationale, but argue that the method-oriented application of quantitative methodologies is nonetheless a valid approach to analysis.

Decision theory, for example, deals with choice in the face of uncertainty. In particular, it deals with the problem facing a decisionmaker who must choose between alternative courses of action in the face of uncertainty about what outcome will result from the action he chooses. It prescribes the way in which the decisionmaker should integrate his uncertainties and his preferences among (or utilities for) outcomes, in order to select the best decision in the light of those uncertainties and preferences. The elements of the decision problem dealt with by decision theory--the decisionmaker's alternatives, uncertainties, outcomes, preferences--are elements that can be found in most problems of policy choice. Hence, decision theory appears to provide a useful methodology for addressing and dealing with such problems. Indeed, many quantitative analyses are of this form--the model used is a quantitative decision problem, and the results of the analysis prescribe the optimal choice within that problem. The literal rationale for such analyses derives from equating the words used in the theory--decisionmaker, uncertainty, preference, etc.--with the same works in the substantive problem. Thus, "uncertainty" in the substantive problem is assumed to be adequately captured by "uncertainty" in the model, the set of alternatives in each are assumed to be the same, etc. It follows logically that the choice prescribed by analysis of the model is optimal in the substantive problem.

In cases where this literal identification of problem and model can be justified, the literal rationale provides strong grounds of support for the analysis and conclusions. These cases, however, are more the exception than the rule, particularly among problems requiring any significant degree of creative insight. In most problems, the validity of literal problem/model identification will be somewhere between questionable and obviously specious. In these cases, the literal rationale is invalid, and if the methodology is to be justified, a different rationale must be found. The pragmatic rationale says, in effect, that even if the methodology doesn't do what is claims, it's still a useful aid to decisionmaking, or to understanding squishy problems. The reasons why this is claimed to be true vary, ranging from "You wouldn't want to trust judgment alone, would you?" to arguments that quantitative analysis helps lay bare the structure of the problem and provides decisionmakers with a better understanding than they would have otherwise had. There are, in effect, different pragmatic rationales applicable in different degrees to different problems. None are supportable or refutable on strictly logical grounds.

The Genie in the Eight Ball: The structure of these rationales can perhaps be seen more clearly if we set quantitative methodology aside for a minute and examine another decision aid or source of knowledge --the Genie in the Eight Ball. Most five-and-dime and magic stores sell a large plastic eight ball with a small window on the bottom, filled with a black liquid. Inside is a polyhedral float with inscriptions such as "yes," "no," "ask me again," "better not," etc., on the faces. To get advice from the eight ball, you phrase a question that can be answered yes or no, turn the ball over, and see what answer floats up to the window. One rationale that might be given for the efficacy of this procedure is that the answer is determined by a Genie who lives in the eight ball, and who talks directly with the gods who control the world.

Now, suppose I have a friend who advocates using the Genie in the Eight Ball to deal with squishy problems. Let's examine the rationales he might put forth to support that procedure, and the parallels that exist in rationales for quantificationism.

My friend's basic rationale might well be that the Genie speaks directly with the gods, and the gods determine how the world works. Asking the Genie is thus justified by the fact that he provides a direct line to the gods. This theological argument corresponds to the literal rationale for quantificationism, in the sense that it is obtained by interpreting the description of the methodology literally. Occasional failures of the Genie to provide good advice do not negate this argument. They can be explained by the fact that the Genie is capricious, that he may not always understand the gods, or that the gods may change their minds after they've talked to him. If my friend truly believes, there is likely to be little I can say to convince him otherwise, and if I am truly skeptical, there will be little he can do to convert me.

The major difference between this theological rationale for the Genie and the literal argument for quantificationism lies in the fact that I can accept or reject the former on a one time problem-independent basis, while the latter requires an *ad hoc* determination based on the nature of the question. In other words, I accept or reject the existence of the Genie independently of the question I asked him to answer for me. I cannot make the same problem-independent decision about the validity of mathematical models of real phenomena, however, since I can find some instances in which models are clearly valid and others in which they are clearly invalid. The question of validity for a particular problem/model pair, therefore, is one that must be dealt with on an *ad hoc* basis. While this complicates the problem of accepting or rejecting quantitative analysis, it does suggest rejection of the literal rationale for quantificationism. Too many counter-examples exist for universal acceptance of that rationale.

But what of the pragmatic? Consider my friend with the eight ball. If I point out to him that I don't believe in the Genie and, hence, think the eight ball is a lousy decisionmaking device, he might respond that he doesn't believe in the Genie either. He might say, in fact, that anyone would be silly to really believe in the Genie, but that regardless of whether the Genie really exists or not, the eight ball is a useful pragmatic decisionmaking tool. Among the reasons that he might cite are the following.

Managerial. He helps make arbitrary decisions. Many of the decisions most people or organizations face are largely arbitrary in the sense that the critical choices have been made in selecting the alternatives. The final choice doesn't really matter much, at least so far as can be told on the basis of current knowledge. The choice of where I go to lunch, for example, is in this category. As long as I am reasonable in the set of alternatives I consider--i.e., I pick places nearby, within my price range, and with reasonably sanitary conditions--my actual choice is irrelevant, and it would be a waste of time to agonize over the decision. If I have trouble making arbitrary decisions and need some sort of rationale for doing so, asking the Genie and accepting his advice can be a very useful procedure.

Empirical. His advice has been good in the past. My friend might argue that the Genie has shown himself to be useful and reliable in the past, as demonstrated by the worth of the advice he has provided. If the decision of where to go for lunch is not recognized as essentially arbitrary, for example, the fact that never once had following the Genie's advice led to food poisoning (or even indigestion) might be cited as evidence of his efficacy. My friend might also be able to cite a few more significant decisions in which the Genie had played a major role. Suppose, for example, that he had recently considered the purchase of either a station wagon or a motorcycle, and on the Genie's advice, had purchased the station wagon. If he has a wife, four kids, and a large dog, and he likes to take weekend family outings, he might point out that the motorcycle would have clearly been an inferior choice. The Genie's advocacy of the station wagon thus appears to provide further evidence of his efficacy and reasonableness.

Catalytic. He jogs the thought process. My friend could also argue that he knows better than to take the Genie's advice without question, but that discussing things with the Genie is a useful catalyst in clarifying issues and in identifying important factors that might otherwise go unnoticed. In effect, he might say that it's not the answer that matters, but the way he asks the questions. His seeking advice on the motorcycle-station wagon decision, for example, may have involved a long and drawn out dialogue on the relative merits of each, rather than a simple yes-no question of "Should I buy a motorcycle?"

Ritualistic. In further support of any of these rationales my friend might argue that the Genie is a useful social convention, a ritual that people use even though they know better, in the same way we anthropomorphize "Mother Nature" or "Jack Frost." He could argue, in effect, "Of course I don't believe in the Genie, but he provides a useful terminology and a useful framework for describing how I do things."

Contingent. The answers the eight ball provide are contingent on the existence of the Genie, and it's up to the user to decide whether he is real or not. Suppose that, in addition to (or perhaps instead of) using the Genie to make his own decisions, my friend makes a living as a fortune teller, providing advice from the eight ball to others. He then need not evaluate his own belief in the Genie. Instead, he might argue "All I do is tell the people what the Genie says, and it's up to them to decide whether they believe in him or not and to accept or reject his advice accordingly. I lay no claim to knowledge, but rather to the ability to turn the eight ball over and read the answer that comes up. What I do is clear, and its up to the customer to decide whether or not it is useful to him."

Unlike the technological argument, these are not rationales that can be universally accepted or rejected on a problem-independent basis. They all have some merit, depending on the question asked, and alternative forms of inquiry available, and other contextual factors. If my friend would find himself in a constant quandry and turmoil over where to eat lunch or what kind of gasoline to buy, for example, the Genie may serve a very useful function in providing him a means of dealing with that type of question. If he feels that the eight ball really has worked in the past, then, in some sense, that justifies using it in the future. This is particularly true if the Genie is in fact a good catalyst to his thinking--if conversations with the Genie truly served to peel back the layers of my friend's consciousness and expose information, ideas, etc., that he might otherwise fail to consider. If he makes his living with the eight ball, then the contingent rationale is particularly cogent. In using that rationale, however, it is important to distinguish between belief in the efficacy of the method and economic self-interest. If my friend wants to truly understand the validity of what he does, he must be careful to avoid overselling himself on the existence of the Genie in order to rationalize his livelihood. If he wants to consider himself an ethical fortune-teller, then it is incumbent on him to exercise a great deal of care in insuring that he doesn't oversell his method to his customers, but that he is as careful in defining it's limits as he is in touting its advantages.

These rationales all have parallels among the rationales for quantificationism. Moreover, because quantitative analysis is a far more complex methodology than the Genie in the Eight Ball, these parallel rationales are correspondingly more complex.

Many of the decisions faced by government, especially at fine grain levels of detail, are of the largely arbitrary type. Final budget allocations, once reasonable limits have been decided on, for example, are in this category, as are most choices between largely similar but competing programs. This is particularly true when a fair margin of uncertainty exists about the precise utility of the program or about the incremental return from additional investment. In such cases, well-defined repeatable procedures to make precise choices among reasonable alternatives are needed. Quantitative procedures satisfy this need very well.

Many of the empirical arguments for quantificationism rest heavily on this managerial justification--"Look how much smoother things have run since we introduced program budgeting," etc. Quantificationist techniques and procedures are often highly efficient as management tools, largely because of the extent to which they simplify problems and provide repeatable mechanical decision procedures. The very characteristics, however, that make any methodology (quantitative methodology or the Genie) a useful managerial device for routine decision-making make it a risky device for catalytic mind jogging. The risk occurs because the terminology used, etc., make it quite easy to attribute validity to the mechanical application of the methodology rather than to its use as a catalyst. Once this is done, it is easy to slip into the habit of applying it mechanically and accepting the answers thus obtained as justified because of the way they were obtained. In spite of the fact that my friend found the Genie a useful device for exploring the relative advantages of the motorcycle and the station wagon, for example, he is deluding himself if he attributes the wisdom of his decision to buy the station wagon to the Genie. In so doing, moreover, he increases the likelihood that he will misuse the Genie next time, by listening to the Genie and ignoring his own good sense when the two conflict.

The ritualistic and contingent rationales are particularly pernicious in this regard, because they encourage sloppiness about just what is being analyzed and just what the analysis shows, and a sense of complacency about the potential impact of that sloppiness. They also serve as "cop outs," relieving the analyst of responsibility for the substantive interpretation of his conclusions, and transferring that responsibility to "them," "everybody," or the customer for or reader of his analysis. The ritualistic rationale takes such forms as "This formulation leaves a lot out, but it is the one everybody uses so it will do for now," or "Everybody uses that terminology and understands its limitations," etc. Contingent rationales occur in justifications such as "It's not up to me to decide if the assumptions are reasonable, but only to insure that the conclusions follow," etc. These rationales seem to assume that the good judgment of all involved is sufficient to overcome the weaknesses and inadequacies of literal interpretations of method-oriented analysis. This assumption is in direct contradiction to that portion of the literal justification which argues that explicit systematic analysis is necessary because of the inadequacies of subjective judgment.

<u>An Image of Analysis</u>. There is a prominent contemporary image of the quantitative analyst as a purveyor of objective fact based on logic and hard data. He takes no personal responsibility for his conclusions, since they are not of his making, but are inherent in the nature of things. All he has done is uncovered them and made them visible for all to see. This is conveyed by the fact that he never talks about what <u>he</u> thinks, but restricts himself to what "the analysis shows" or "is clear from the data." In the sense that he is

perceived as not personally involved with his conclusions, he is like the natural scientist, or perhaps the priest who serves only as a conduit to the gods. He distains the "merely qualitative" and often speaks perjoratively of "subjective judgment."

Now very few people (there are some) "really believe" in this image, in the sense of accepting it without qualification after careful reflection. This is the image, nonetheless, that serves to define the operational role of the quantitative policy analyst in contemporary society. He pretends that it is true, and his customers pretend that it is true, even though all know better.

This results in a set of conventions that define how the analyst will perform and report on his analysis, and how the customer will react to it. The analyst chooses a model he knows how to analyze, analyzes it as though it were an adequate representation of his problem, and interprets his results accordingly. He reports these conclusions and describes enough of the model and the mathematical analysis to show that he performed the ritual correctly. This description usually contains a listing of "assumptions" from which the conclusions are said to follow. It frequently contains the disclaimer that the evaluation of the assumptions is the responsibility of the customer, not the analyst. It may also contain the disclaimer that the analysis addresses only a part of the problem with which the customer has to deal, and additional factors must be taken into account by the customer in acting on the analysis. In spite of these disclaimers, however, the conclusions are frequently stated in positive and unequivocal terms.

These conventions allow for the production of a great deal of good analysis, since they give competent and insightful analysts much leeway within which to operate. They also allow the production of much bad analysis, since they focus attention primarily on the technical correctness of the mathematical analysis rather than on its substantive validity or relevance. This, in turn, encourages sloppy method-oriented analysis of little substantive value.

The government policy- or decisionmaker who is the customer for the analysis also has a role to play in support of this image. He pretends he believes in the image, while at the same time, independently judging the worth of the analysis presented to him. He must reject the bad and use (and perhaps even learn from) the good. The hedges provided by the analyst, in the form of assumptions the customer can disagree with or "other factors" not included in the analysis that he must consider, allow him to do this without seriously questioning or coming into direct conflict with the image of the analyst as a giver of fact.

Everyone goes along with the charade because everyone else does, and it seems to be a relatively harmless charade as long as everyone understands it.

But is it? That itself is a very squishy question. It involves the assessment of how well people really perform when doing one thing while pretending to do another, neither of which is well defined in any formal objective sense. There is no objective way, really, to decide. A number of arguments, however, suggest that we would get more out of our policy analysis if we were more honest about the difficulties of the task, and about the importance of careful subjective judgment in coming to grips with and coping with those difficulties.

There is no way of really being sure how well people really understand the nature of the charade. The amount of bad analysis done by seemingly competent and sincere analysts would suggest that they don't fully understand it. At the same time, some of the arguments made by policymakers that seem to be based on or swayed by bad analysis would suggest that they don't all really understand it fully either.

Even if they do, role-playing gets in the way. It tends to encourage bad or sloppy analysis by analysts who should, and do, know better, because that's what they see the system as expecting of them. It tends to encourage sloppy evaluation and use by the customers of analysis, particularly when the analysis appears to support a position toward which they feel well disposed.

The charade discourages thoughtful systematic consideration of non-quantitative factors. It does this by defining the ground rules for debate--the terms in which issues are to be framed and decided. It directs attention and effort to the quantitative aspects of basically unquantifiable problems and tends to force debate into that mold. It becomes, somehow, too inconvenient, too cumbersome, etc., to systematically address the aspects of the question that don't fit in the quantitative mold.

Over time, this introduces institutional distortions, in the sense that people begin to accept simplistic characterizations of problems as the institutional way of looking at them, even though they, individually, know better. What happens eventually, then, is that the institution as a whole settles for and operates with characterizations of its problems that most people in it know is inadequate.[6]

All of these factors combine to suppress the positive potential that systematic analysis has to provide in understanding and coping with complex and squishy policy problems. On balance, then, the contention that the charade is harmless is questionable. It has a positive potential for harm and this potential seems sufficiently great to make efforts to find alternatives worthwhile.

Objectivity and Subjectivity

One of the underlying imperatives toward quantificationism is a desire for "objectivity," without an adequate understanding of what "objectivity" means in the context of analysis of very squishy problems and of the tradeoffs involved in achieving different kinds of objectivity. The natural sciences are seen as objective--as unbiased, quantitative, logical, and free-standing in the sense discussed earlier. These characteristics are equated with objectivity, and the positive valuation attached to them as attributes of science is transferred to the characteristics themselves. Simultaneously, the converse characteristics--biased, qualitative, intuitive, and judgmental --are denigrated and labeled "subjective." "Objectivity" is seen as good, desirable, something to be emulated while "subjectivity" is viewed as negative, perjorative, something to be avoided. This labeling identifies qualities (such as "quantitative" and "unbiased") as always occurring together, and hence, equivalent. They do seem to occur together in the natural sciences, but this will not necessarily be true in the kinds of squishy problems encountered in policy analysis.

It is important, therefore, to distinguish between different meanings of "objective" and to identify situations in which those meanings may coincide or be contradictory. When different forms of objectivity are not simultaneously achievable, we must choose between them. In some circumstances, the most "objective" analysis may be analysis grounded in a high degree of carefully considered subjective judgment.

The term "objective" has two distinct meanings relative to analysis and the conclusions produced by analysis. The first is what I've referred to earlier as "free-standing"--detached from the analyst and residing in a literal statement of the conclusion and its grounds. Objectivity, in the free-standing sense, is a characteristic of a conclusion, or product of analysis, and occurs when the conclusion is grounded entirely in logic and intersubjectively verifiable fact. A second and distinct meaning of "objective" is "unprejudiced"--coming to the problem without preconceived ideas about what the conclusion should be, trying to see things as they really are rather than as the analyst would like them to be, etc. Objectivity in the unprejudiced sense is more a characteristic of the analyst and of the process of analysis than of the conclusion per se.

Corresponding to these two forms of objectivity are complementary forms of subjectivity--judgmental and prejudiced. Subjectivity, in the judgmental sense, is a property of the conclusion and the grounds --the property that the conclusion remains grounded, at least in part, in the subjective knowledge and judgment of the analyst. The grounds reside with the analyst in ways not fully articulated in the language describing them, or fully representable in terms of logic and objective fact. Subjectivity, in the prejudiced sense, is a property of the analyst or the process of analysis he brings to bear. It means a coming to the problem with a preconceived bias about the nature of the problem or about the conclusion to be reached.

The logical compatibilities between these definitions are shown in Figure 5. "Free-standing" and "judgmental" are mutually exclusive properties. Hence, they are logically incompatible. Similarly, "prejudiced" and "unprejudiced" are mutually exclusive and incompatible.

Objectivity	Subjectivity	
	Judgmental	Prejudiced
Free-standing	Incompatible	Compatible
Unprejudiced	Compatible	Incompatible

FIGURE 5. Logical Compatibilities between Forms of Objectivity and Subjectivity

There is, however, no logical incompatibility between a prejudiced analyst and free-standing knowledge or between an unprejudiced analyst and a judgmental conclusion.

These two forms of objectivity seem to coincide in the natural sciences--unprejudiced inquiry leads to free-standing knowledge. This coincidence is a property of a class of problems that the natural sciences addresses, however. It is not one that we have any strong a priori reason to expect in other areas of inquiry or analysis. To insist on the production of conclusions that superficially appear free-standing without first determining that the conclusion sought can be reasonably expected to have that property is itself a severe form of prejudice. What it is likely to produce (and often does in method-oriented quantitative analysis of squishy problems) is superficially free-standing pseudo-knowledge of questionable relationship to the substantive problem. This pseudo-knowledge may be useful, but if it is, it is only through careful judgmental interpretation, in combination with additional judgmental knowledge of the problem possessed by the interpreter.

Much as we might wish it so, there is no a priori guarantee that the two meanings of objectivity will necessarily coincide in policy analysis. Indeed, there is good reason to expect that they will not, since many of the problems encountered possess a political/social/behavioral content of the type man seems best able to understand and cope with judgmentally. In such problems, we cannot "see things as they really are" while restricting our attention to the quantifiable or analytically tractable aspect of the problem. We cannot, therefore, have both forms of objectivity, and we must choose between them. I believe the choice should be in favor of unprejudiced objectivity--of attempting to understand the problems as thoroughly as possible--acknowledging that this means acceptance of judgmentally grounded conclusions in many cases. To insist on the superficial appearance of free-standing objectivity--as the quantificationist does--is itself a severe form of subjective prejudice that can significantly distort the analysis and the conclusions reached.

Again, this is not to imply that quantitative methods have no role in the analysis of squishy problems. They do. That role should be as an adjunct to and in support of carefully considered subjective judgment, however, rather than as an attempted replacement for judgment. For reasons outlined earlier, current theories from which quantitative methodology derive provide inadequate guidance as to how this might be done.

FOOTNOTES

[1]In this regard there is a significant difference worth noting between mathematicians engaged in mathematics, per se, and analysts engaged in the study of substantive problems in the way they relate mathematical models to the real world. The analyst concerned with the real world uses the model as an aid to understanding that world. He, thus, chooses his model to reflect relevant aspects of reality, and when the two diverge, will be likely to modify the model to correspond more closely with the appropriate reality. The mathematician studying the behavior of the abstract model, on the other hand, may be more interested in the behavior of the model than in the real world process he uses as a heuristic device to help him to identify with the model. When he talks of "tossing a coin," for example, as a heuristic description of a sequence of Bernoulli variables, his real interest is likely to be in the behavior of those abstract random variables--not the behavior of real coins.

[2]The terms "substantive problem" and "conclusion" or "substantive conclusion" will be used to refer to things of interest in the real world, and "model" and "results" to the corresponding entities at the mathematical level of analysis.

[3]For additional discussion of this point, see Strauch (1971).

The reader who is so inclined should have no trouble quibbling with the details of my characterization of quantificationism. I could easily do so myself. The relevant question, however, is whether a doctrine like quantificationism is operative in our society today--not whether or not I have managed to define it precisely. A colleague once suggested that quantificationism must be a strawman, because no competent analyst would defend quantificationism as I have defined it, or would admit to the practice. That argument assumes a consistency of belief and action that few men achieve. It is hard to find a defender of hypocrisy or self-deception, either. Nonetheless, we all engage in them on occasion.

[5]Reductionism, in this context, refers to the thinking style that treats a phenomenon of interest as an isolated system, develops a quantitative model for that isolated system, and uses that model as a surrogate for the phenomenon in analyzing or answering questions about the phenomenon.

[6]The effects of this phenomenon in strategic defense planning are discussed in Strauch (1972).

REFERENCES

[1] Strauch, Ralph E., The Fallacy of Causal Analysis, The Rand Corporation, P-4618, April 1971.

[2] Strauch, Ralph E., Winners and losers: a conceptual barrier in our strategic thinking, Air University Review 23 (July-August, 1972), 33-44. Also available as P-4769, The Rand Corporation, June 1972.

MATHEMATICS OF CONFLICT, Martin Shubik (ed.)

DUELS: AN OVERVIEW

George Kimeldorf

Programs in Mathematical Sciences
University of Texas at Dallas
Dallas, Texas

1. INTRODUCTION AND SUMMARY

A dual is a mathematical model for conflict environments in which two competitors can repeat an action several times in an attempt to achieve a certain objective. The strategic question is to determine the optimal times to take the actions in order to be the first to achieve the objective. The later an action is taken, the greater is its chance of success; on the other hand, a competitor must not delay too long lest his opponent act and succeed first. For example, two nations may be competing for the first successful exploration of a celestial body. Country A can afford a maximum of m probes and Country B can afford a maximum of n probes. Each probe may or may not succeed, but, because of constantly improving telemetry technology, the later any given probe is fired, the better is its chance for success. The problem is to determine the optimal schedule for launching the probes. Other examples may involve the firing of missiles, the development of military devices, the launching of information-retrieval satellites, or the making of concessions at a disarmament conference.

The traditional model for the situations we propose to study is the battlefield, where two combatants are each armed with a certain number of bullets and slowly advance toward each other until one kills the other. Because of this model, such situations have been referred to in the literature as "duels" and the actions referred to as firing "bullets." For simplicity, we shall adopt this terminology in this paper but it should be pointed out that the applicability of the theory is much more general.

Without loss of generality, we assume the duel begins at time $t = 0$ when Player 1 is armed with m bullets, and Player 2 with n bullets, each of which may be fired at any time. If Player 1 or Player 2 fires a bullet at any time t, he succeeds (e.g. kills his opponent) with probability $P_1(t)$ or $P_2(t)$, respectively. The functions P_1 and P_2 are called the accuracy functions for the respective players and are continuous and nondecreasing. It is assumed that the accuracy functions are initially 0 when $t = 0$ and eventually increase to 1 when $t = 1$. The numbers m and n of bullets and the accuracy functions P_1 and P_2 are assumed known to both players.

Mathematically, a duel is a zero-sum two-person game, where the first player to succeed wins the game and gets a payoff of +1 from his opponent. In case of a tie the payoff is zero. The value v of a duel is the value as defined in the game-theoretic sense: the probability that Player 1 succeeds first minus the probability that

Player 2 succeeds first, so that $-1 \le v \le 1$.

A bullet can be one of two types: noisy or silent. A bullet is noisy when its firing can be detected by the opponent; a bullet is silent if the opponent does not know it has been fired and hence cannot modify his response accordingly. A noisy duel is a duel in which all bullets are noisy, and a silent duel is one in which all bullets are silent. Hence in a noisy duel, each player knows how many bullets he and his opponent have at every instant of time.

This model can be extended in many directions, such as by allowing continuous firing, permitting the numbers of bullets available to be random, allowing semi-noisy bullets, etc. These and other extensions of the model will be discussed in Sections 2 and 3.

Example 1

For example, consider a noisy duel in which $m = n = 1$. An optimal strategy for Player 1 is of the form: Plan to fire at time t_1 $(0 < t_1 < 1)$, but if Player 2 fires before t_1 , then withhold fire until time 1 , when a hit would be certain. If player 2 adopts a similar strategy (using t_2 for his planned firing time), the expected payoff (to Player 1) is

$$\begin{aligned} K(t_1, t_2) &= 2P_1(t_1) - 1 && \text{if } t_1 < t_2 \\ &= P_1(t_1) - P_2(t_2) && \text{if } t_1 = t_2 \\ &= 1 - 2P_2(t_2) && \text{if } t_2 > t_1 \end{aligned}$$

It thus follows that an optimal strategy for either player is to plan to fire the bullet at a time t_0 for which $P_1(t_0) + P_2(t_0) = 1$, and wait until time 1 if the opponent fires first and misses. The value of this duel is $P_1(t_0) - P_2(t_0)$.

Other elementary examples are discussed in Dresher (1961) and Karlin (1959).

Section 2 below presents a survey of the background of the theory of duels, with particular emphasis on the more recent advances in the area, and also outlines the present status of knowledge. Some open problems are discussed in Section 3. Not included in this overview is the somewhat distinct but important subject called "stochastic duels," in which there exists a large and rapidly expanding literature.

2. BACKGROUND AND PRESENT STATUS OF KNOWLEDGE

The game-theoretic study of duels began in 1948 at the RAND Corporation. In general, the early papers solved individual duels in which the players each had 1 or at most 2 bullets. Among the early contributors to the theory of duels were R. Bellman, R. L. Belzer, D. Blackwell, M. A. Girshick, I. Glicksberg, L. S. Shapley, et al. Shiffman (1953) considered the general symmetrical duel with one bullet for each player and Karlin (1953) related the general solutions of duels with 1 bullet each to the theory of positive integral transforms.

The solutions of certain duels with arbitrary (finite) numbers of bullets for each player appear in Blackwell and Girshick (1954). In particular, they solved the special class of noisy duels in which the accuracy functions P_1 and P_2 are assumed equal. If Player 1 has m bullets and Player 2 n bullets with $P_1 = P_2$, the solution can be described recursively as follows: Let t_{mn} be a solution to $P_1(t_{mn}) = P_2(t_{mn}) = 1/(m+n)$. If $m \geq n$, it is optimal for Player 1 to fire his first bullet at time t_{mn}; if he misses and Player 2 does not also fire at time t_{mn}, Player 1 should proceed as if the players originally had $m-1$ and n bullets, respectively; if Player 1 misses at time t_{mn} and Player 2 also fires and misses at time t_{mn}, Player 1 should proceed as if the players originally had $m-1$ and $n-1$ bullets, respectively. If $m < n$, Player 1 should wait until time t_{mn} to see whether Player 2 fires at that time (which he should do); if Player 2 fires at time t_{mn} and misses, Player 1 should proceed as if the players originally had m and $n-1$ bullets, respectively; if Player 2 fails to fire at time t_{mn}, Player 1 should fire immediately thereafter (more precisely, Player 1 would choose $\varepsilon > 0$ and fire at random at some time in the interval $(t_{mn}, t_{mn}+\varepsilon)$). The value of the game is $v = (m-n)/(m+n)$. Thus, optimal strategies do not in general exist, but Player 1 has for any $\varepsilon > 0$ a strategy, called an ε-optimal strategy, which guarantees him an expected payoff of at least $v-\varepsilon$, while Player 2 can guarantee that Player 1 cannot win on the average more than $v+\varepsilon$.

Shortly, thereafter, Restrepo (1957) solved the class of duels with arbitrary (finite) numbers of silent bullets for each player. He assumed the accuracy functions, which need not be equal, to be strictly increasing and continuously differentiable. An optimal strategy for Player 1 in this duel takes the following form: There exist m positive numbers $a_1 < a_2 < \ldots < a_m < 1$. On each interval (a_i, a_{i+1}) for $i = 1, 2, \ldots, m-1$, the i-th bullet is fired (assuming both duelists are then alive) at random according to a probability density function. Thus, the planned firing time for any bullet is independent of the firing times for the other bullets. The m-th bullet is with some probability α fired at time 1 or with probability $1-\alpha$ fired randomly in the interval $(a_m, 1)$ according to a probability density function. The numbers $\{a_i\}$ and the value v of this game, which are uniquely determined by the functions P_1, P_2 and the numbers m, n, cannot be solved simply as explicit functions of these quantities. Thus, in silent duels, each player has an optimal strategy, which involves randomization.

Example 2

For example, consider a silent duel with $P_1(t) = P_2(t) = t$, $m = 2$, $n = 1$. The value of this game is $v = (3-2a)/(2+3a)$, where $a = [1+(1/3)^{1/2}]^{1/2}$. An optimal strategy for Player 1 specifies firing his first bullet in the interval between $a_1 = (1+2a)^{-1}$ and $a_2 = [1+2(1/3)^{1/2}]^{-1}$ at random using the probability density function $3(4+6a)^{-1}t^{-3}$. His second bullet should be fired according to a probability distribution which has density $(3-3^{3/2}/2)t^{-3}$ on $(a_2, 1)$ and has mass $2-3^{1/2}$ at $t=1$. An optimal strategy for Player 2 is to fire his bullet at random in the interval $(a_1, 1)$ with probability density function g defined by

$$g(t) = \frac{2a_1^2 a_2}{(a_1^2 + a_2)} t^{-3} \qquad \text{if} \quad a_1 < t < a_2$$

$$= \frac{2a_1^2 a_2}{(a_1^2 + a_2)(1-a_2)} t^{-3} \qquad \text{if} \quad a_2 \leq t < 1$$

Player 2's strategy assigns no mass at $t=1$. This example is discussed in Chapter 9 of Dresher (1961) and in Chapter 6 of Karlin (1959).

The status of the game-theoretic study of duels as of the end of the 1950's can be summarized as follows: The solution of noisy duels was known only for the special cases when $P_1 = P_2$ or where $m = n = 1$. The solution of silent duels was known for arbitrary, P_1 , P_2 , m , n . Essentially no progress had been made in solving mixed duels, those in which some bullets are silent and others noisy; only the case $m = n = 1$ had been solved.

Smith (1967) considered mixed duels in which Player 1 has two bullets of which one is silent and the other noisy, while Player 2 has only a noisy bullet. He solved these duels, with arbitrary accuracy functions P_1 and P_2 having positive derivatives, assuming the additional restriction that Player 1 be required to fire his silent bullet first. He showed an optimal strategy for Player 1 to consist of an absolutely continuous probability distribution over an interval (a, t_0) for firing his silent bullet, followed by firing his noisy bullet at the fixed point t_0 . An optimal strategy for Player 2 is to fire his noisy bullet according to a probability distribution which has an absolutely continuous part on (a, t_0) and assigns positive probability to the point t_0 . The point t_0 , as in the noisy duel with one bullet each, is the solution to the equation $P_1(t) + P_2(t) = 1$.

In 1969, Fox and Kimeldorf solved the class of noisy duels with arbitrary numbers m , n of bullets and arbitrary continuous, nondecreasing accuracy functions P_1 , P_2 . Letting $G_{m,n}$ be such a duel for fixed P_1 , P_2 , they showed that $G_{m,n}$ could be solved recursively by using the solution of the duels $G_{m,n-1}$ and $G_{m-1,n}$ In particular, for given accuracy functions P_1 and P_2 , it was shown that there exist a set $\{t_{ij} : i, j = 1, 2, \ldots\}$ and a unique set $\{v_{ij} : i, j = 1, 2, \ldots\}$ of numbers for which

(2.1) $$v_{ij} = P_1(t_{ij}) + [1 - P_1(t_{ij})]v_{i-1,j}$$
$$= -P_2(t_{ij}) + [1 - P_2(t_{ij})]v_{i,j-1}$$

where $v_{io} = 1$ for $i > 0$ and $v_{oj} = -1$ for $j > 0$. For any $m > 0$, $n > 0$, the duel $G_{m,n}$ has value v_{mn}. The times $\{t_{ij}\}$ can be found from the equations

(2.2) $$\prod_{i=1}^{m} [1 - P_1(t_{1n})] + \prod_{j=1}^{n} [1 - P_2(t_{mj})] = 1$$

and can be interpreted as follows: If both Players play optimally, then the first shot in $G_{m,n}$ will be fired at time t_{mn}. It will be fired by Player 1 if

(2.3) $$P_1(t_{mn}) - P_2(t_{mn}) + [1 - P_1(t_{mn})][1 - P_2(t_{mn})]v_{m-1,n-1} \geq v_{mn}$$

and by Player 2 if the inequality (2.3) is reversed. Thereafter, the Players follow their optimal strategy in the reduced game $G_{m-1,n}$, $G_{m,n-1}$ or $G_{m-1,n-1}$. Loosely speaking, the first shot is fired by a player for whom simultaneous firing at time t_{mn} is not disadvantageous. Thus, a proper strategy for Player 1 can be described recursively as follows: If (2.3) holds it is optimal for Player 1 to fire his first bullet at time t_{mn}; if he misses he should thereafter proceed as in $G_{m-1,n}$ or $G_{m-1,n-1}$ depending on whether or not Player 2 also fires at time t_{mn}. If (2.3) does not hold, then Player 1 should wait until time t_{mn} to see whether Player 2 fires at that time (which he should do); if Player 2 fires at time t_{mn} and misses, Player 1 should proceed as in $G_{m,n-1}$; if Player 2 fails to fire at time t_{mn}, Player 1 should fire immediately thereafter (more precisely, Player 1 would choose $\varepsilon > 0$ and fire at random in the interval $(t_{mn}, t_{mn} + \varepsilon)$). An analogous strategy with (2.3) reversed would be appropriate for Player 2.

Thus, optimal strategies do not in general exist, although for every $\varepsilon > 0$, each player has an ε-optimal strategy which may involve randomization. If both players play optimally, no randomization will in fact occur and all shots will be fired at times t_{ij} given by (2.2). Randomization occurs only in response to an opponent's failure to fire at the prescribed time.

Note that the ε-optimal strategies defined recursively above only make sense if

(2.4) $$t_{mn} < \min(t_{m-1,n}, t_{m,n-1})$$

In fact, the proof by Fox and Kimeldorf (1969) of the validity of the above solution rests on showing that the t_{ij} defined by (2.1) or (2.2) also satisfy (2.4)

Fox and Kimeldorf (1970) expanded on their 1969 paper by presenting a detailed discussion of criteria for the existence of good first-shot times and by displaying tables, which in many cases could be used to trace the play of the game. Suppose we change the rules of the game so as to permit firing only at fixed times $0, 1/k, 2/k, \ldots, 1$, with the further stipulation that each of these times is labeled (perhaps in a way which depends on the previous actions of the players) either as a time when only Player 1 may shoot or only Player 2 may shoot, thus prohibiting simultaneous firing. This modified noisy duel then becomes a perfect information game. It is well-known that such games have values, and nonrandomized optimal (as opposed to ε-optimal) strategies exist for both players. A reason that the original game G_{mn} fails to be a perfect information game is the possibility of simultaneous firing: a player must make his decision of whether or not to fire at time t before he knows whether his opponent also plans to fire at time t. To take into account the possibility of the two players' firing simultaneously, we set $v_{oo} = 0$ and denote the left hand side of (2.3) by b_{mn}, which is the expected payoff within ε, if both players in G_{mn} fire their first shots at time t_{mn} and follow ε-optimal strategies thereafter in $G_{m-1,n-1}$. Clearly, if $v_{mn} \le b_{mn}$, Player 1 does not mind if simultaneous shooting occurs at time t_{mn} and can therefore fire at that time. On the other hand, if $v_{mn} > b_{mn}$ Player 1 does not want such simultaneous firing to occur. He can make the probability of simultaneous firing 0 by planning to fire not at time t_{mn}, but at a time chosen at random from a small interval following t_{mn}.

Fox and Kimeldorf also showed that computing the values or strategies in any noisy or silent duel with accuracy functions $P_1(t)$ and $P_2(t)$ is equivalent to computing the values or strategies in the duel with accuracy functions t and $P_2(P_1^{-1}(t))$. Thus, without loss of generality, we can always assume that $P_1(t) = t$.

Example 3

Consider the noisy duel in which $P_1(t) = t$, $P_2(t) = t^2$, $m = 2$, $n = 4$. If both players proceed optimally, the first bullet would be fired at time .301 by Player 2; if the players survived that long, successive firing would occur at time .334 by Player 2, at time .382 by Player 1, at time .477 by Player 2, and finally at time .618 by both players simultaneously. The value of this duel is .093, which means that Player 1, who had fewer bullets, would have a slight advantage due to his superior accuracy.

An example, such as this, in which the player with fewer bullets could overcome his disadvantage through increased accuracy led Fox and Kimeldorf to investigate "the price of inaccuracy." Specifically suppose Player 1 has fixed accuracy function P_1 and n bullets while Player 2 has fixed accuracy function P_2. Asymptotically, how many bullets m as a function of n for large n does Player 2 need to ensure that the value of the game is not greater than 0? This is a realistic question in the sense that a player involved in a duel might need to decide how to best allocate his resources: should he improve his accuracy or increase his amount of ammunition.

This question of relative cost effectiveness and optimal allocation of resources has obvious military applicability. Unfortunately, this question could not be answered for either silent duels or noisy duels, because the formulas for the value are complicated implicit functions of m, n, P_1 and P_2. This question was answered eight years later when Kimeldorf and Lang (1978) proved that if $P_1(t) = t$ and $P_2(t) = ct^r + h(t)$, where $h'(t) = o(t^{r-1})$ as $t \to 0$ for $r \geq 1$, then m is proportional to n^r. More precisely, it was shown that

$$\text{(2.5)} \qquad \lim_{n\to\infty} mn^{-r} = [d(r)]^{1-r} r^{-r} c^{-1}$$

where

$$\text{(2.6)} \qquad d(r) = \int_1^\infty u^{-2}(1 + u^{r-1})^{-1} du$$

Surprisingly, this result is valid regardless of whether the duel is silent, noisy, or mixed. The details of this paper and the role of noise will be discussed further below.

Silent duels allowing continuous firing were discussed by Yanovskaya (1969). In these duels, each competitor has a fixed amount of ammunition which can be discharged with varying intensities of fire. Lang and Kimeldorf (1975) proposed an improved mathematical model for silent duels with continuous firing, called machine-gun duels, and solved some special cases. To motivate the model for a continuous-fire duel consider a silent discrete-fire duel with n bullets each and assume for the moment that Player 2 is not shooting, so that Player 1 is essentially shooting at a target. If Player 1 chooses n times t_k, where $0 \leq t_1 \leq t_2 \leq \ldots \leq t_n \leq 1$, at each of which he shoots one bullet, we define $S(t_1, t_2, \ldots, t_n)$ to be the probability that he hits his target (at least once).

In a duel with continuous firing, a pure strategy for Player 1 is an absolutely continuous, nondecreasing function F on $[0,1]$ satisfying $F(0) = 0$ and $F(1) \leq 1$. We call F a <u>cumulative intensity function</u> and its derivative f an <u>intensity function</u>. There is again assumed to be a function $S(f)$ representing the probability that Player 1 hits his opponent if he chooses to fire in $[0,1]$ according to f and Player 2 is not firing. Our problem is to determine what S should be in order to make this a natural generalization of a discrete-fire duel.

For any intensity function f, let $S_I(f)$ denote the probability that Player 1 hits the target in the interval I using intensity function f. For fixed $t \in [0,1]$, consider the intensity functions

$$f_{t,n}(s) = n \quad \text{if} \quad t < s < t + 1/n ,$$
$$= 0 \quad \text{otherwise}$$

which, as n gets large, concentrate more and more firepower at time t. Then, for $0 \leq t < 1$, assume $\lim_{n\to\infty} S_{[0,1]}(f_{t,n})$ exists. Call this limit $P(t)$ and assume further that $P(0) = 0$,

$\lim_{t\to 1^-} P(t) = 1$, and that P' exists and is continuous and positive on (0,1). Loosely speaking, $P(t)$ represents the probability that he hits his target at time t if he shoots one unit of ammunition at this time.

Let $P_T\{f,P,t,h\}$ be the probability that Player 1, using intensity function f with accuracy function P , hits his target in the interval $(t, t+h)$. As in previous models, assume that there exists a function H , depending on P , such that

$$P_T\{f,P,t,h\} = f(t)H(t)h + o(h) \tag{2.7}$$

It follows easily from (2.7) that

$$S_I(f) = 1 - \exp\left[-\int_I H(s)f(s)ds\right]$$

for any subinterval I of (0,1). In particular, for a given $t \in [0,1]$,

$$S_{(0,1)}(f_{t,n}) = 1 - \exp\left[-n\int_t^{t+1/n} H(s)ds\right] \tag{2.8}$$

Since by assumption the left-hand side of (2.8) has a limit $P(t)$.

$$\lim\left[1 - \exp\left(-n\int_t^{t+1/n} H(s)ds\right)\right] = 1 - \exp[-H(t)] = P(t)$$

and hence

$$H(t) = -\log[1 - P(t)]$$

Thus, Lang and Kimeldorf heuristically motivated adopting this model for a silent duel with continuous firing. That is, for the i-th player

$$P_T\{f_i, P_i, t, h\} = f_i(t)A_i(t)h + o(h) , \tag{2.9}$$

where $A_i(t) = \log[1 - P_i(t)]$. From (2.9) the payoff-kernel

$$K[f_1, f_2] = \int_0^1 \{f_1(t)A_1(t) - f_2(t)A_2(t)\} \cdot \exp\left[-\int_0^t f_1(s)A_1(s)ds - \int_0^t f_2(s)A_2(s)ds\right]dt \tag{2.10}$$

can be derived in a straightforward manner.

Thus, the model for a silent duel with continuous firing introduced by Lang and Kimeldorf (1975) can be stated formally as follows: For $i = 1, 2$, associated with Player i is a nonnegative, strictly increasing absolutely continuous function P_i defined on [0,1] with $P_i(0) = 0$ and $P_i(1) = 1$. The function P_i , Player i's accuracy function, is known beforehand to both players. There are fixed nonnegative numbers M_1 and M_2 (also known beforehand to both players), representing the amounts of ammunition available to Players 1 and 2, respectively. A pure strategy for Player i is a nondecreasing, absolutely continuous function F_i (with derivative f_i) on $[0,1]$ satisfying $F_i(0) = 0$ and $F_i(1) \leq M_i$. The functions f_i and F_i are called intensity functions and cumulative intensity functions, respectively. The payoff-kernel is given by (2.10), where $f_i = F_i'$.

Lang and Kimeldorf solved this duel in the symmetric case in which $M_1 = M_2$ and $P_1 = P_2$ (so that without loss of generality we can assume $P_1(t) = P_2(t) = t$). The value of the duel is, of course, 0; an optimal strategy for either player is to choose the cumulative intensity function F defined by

$$
\begin{aligned}
F(t) &= 0 && \text{if } t < a, \\
&= 1 + \frac{1}{2\log(1-t)} && \text{if } a \leq t < 1, \\
&= 1 && \text{if } t = 1
\end{aligned}
$$

where $a = 1 - \exp[-1/2]$.

A year later, Lang and Kimeldorf (1976) solved this duel in its greatest generality by dropping the restrictions that $M_1 = M_2$ and $P_1 = P_2$. Define the functions $f_{\lambda a}$ and $g_{\lambda a}$ by

$$
(2.11) \qquad f_{\lambda a}(t) = \frac{A_2'(t) I_{(a,1)}(t)}{A_2(t)[A_1(t) + A_2(t)/\lambda]}
$$

$$
(2.12) \qquad g_{\lambda a}(t) = \frac{A_1'(t) I_{(a,1)}(t)}{A_1(t)[\lambda A_1(t) + A_2(t)]}
$$

where λ and a are as yet unspecified constants in $(0,\lambda)$ and $(0,1)$, respectively, and where for any real numbers $x < y$ $I_{(x,y)}(t)$, the indicator function of the interval (x,y), is defined to be 1 if $x < t < y$ and 0 otherwise. Kimeldorf and Lang then showed that there exist a pair (λ, a), where $\lambda > 0$ and $0 < a < 1$, such that $\int_a^1 f_{\lambda a} = M_1$ and $\int_a^1 g_{\lambda a} = M_2$. The pure strategies f and g given by (2.11) and (2.12) were shown to be optimal strategies for the respective players and the value was shown to be $[\lambda A_1(a) - A_2(a)]/[\lambda A_1(a) + A_2(a)]$.

In the same paper, Lang and Kimeldorf extended the duel further by dropping the restriction that the cumulative intensity functions be absolutely continuous, thus allowing continuous firing or discrete firing as special cases. The payoff kernel in this extended duel is given by

$$
(2.13) \qquad K(F_1, F_2) = \int_0^1 \overline{Q}_2 dQ_1 - \int_0^1 \overline{Q}_1 dQ_2
$$

where $\overline{Q}_i(t) = \exp\left[-\int_0^t A_i dF_i\right]$ and $Q_i = 1 - \overline{Q}_i$. It is easy to verify that (2.13) reduces to (2.10) if the functions F_i are absolutely continuous with respective derivatives f_i. The solution of this extended duel (called a silent duel with nondiscrete firing) is the same as that in which the intensity functions are restricted to be absolutely continuous: Player 1 should use the intensity function $F_1(t) = \int_0^t f_{\lambda a}$ and Player 2 should use the intensity function $F_2(t) = \int_0^t g_{\lambda a}$, where $f_{\lambda a}$ and $g_{\lambda a}$ are given by (2.11) and (2.12) respectively.

Note that the optimal strategies in these silent duels with nondiscrete firing are pure strategies, while no pure strategy can be optimal in discrete-fire silent duels. To see why, note that if a player's pure strategy prescribed firing his first bullet at some fixed time t, his opponent, because of the discontinuity of K, could gain an advantage by firing "just before" t. This intuitive argument for the nonexistence of pure optimal strategies does not apply to silent duels with nondiscrete firing. These duels allow pure strategies which prescribe expending zero units of firepower at any fixed time t. Thus it is not surprising that silent duels with nondiscrete firing have optimal strategies which are pure.

The existence of pure optimal strategies in a non-discrete duel is of considerable interest and does not appear to follow from any standard existence theorem. The existence of such pure optimal strategies lends insight to the role that noise plays in duels. The pure strategy spaces of noisy discrete duels are considerably more complex than those of silent discrete duels. In a noisy discrete duel a player's actions in $(t,1)$ can depend on both his own firing times in $[0,t]$ and on his opponent's detected firing times in $[0,t]$. It is possible to define a noisy non-discrete duel by generalizing a noisy discrete duel in much the same way as the silent non-discrete duel generalizes the silent discrete duel. The strategy spaces in silent non-discrete duels would be subspaces of the strategy spaces of noisy non-discrete duels. That is, any silent duel strategy is a noisy duel strategy that ignores the opponent's observed firing patterns.

Consider now the following property of any optimal strategy for a player in a game. Even if Player 2 knows Player 1's optimal strategy in advance, Player 2 cannot prevent Player 1 from attaining, through the use of his optimal strategy, the value of the game. Notice that knowing a pure strategy in advance is at least as valuable to a player as is being able to 'hear' a pure strategy. However, in a silent non-discrete duel, Player 2 can reveal his pure optimal strategy, and hence give up any advantage of silent firing, and yet still receive the value of the game.

The foregoing argument, which was presented by Kimeldorf and Lang (1977), shows that noise is not a factor in non-discrete duels in the sense that the pure optimal strategies given above for silent non-discrete duels are also pure optimal strategies for the corresponding noisy and mixed non-discrete duels. Of course, the values of these games are also the same.

We see that the non-discrete duel is actually the more natural game to look at and that the classical discrete duels arise when the strategies a player may use are restricted. Specifically, in silent discrete duels, the only permissible pure strategies for a player are those measures that, for a fixed n, place mass $1/n$ at each of n times. In noisy discrete duels, a player is restricted to using as pure strategies those which place mass $1/n$ at each of n times (which may depend on the other player's strategy). The importance of such factors as noise and randomization in discrete games arises from these restrictions.

Asymptotic properties of non-discrete duels were studied by Kimeldorf and Lang (1977). They considered a duel (of any kind) where Player i has modified accuracy function A_i (or corresponding accuracy function P_i) and amount of ammunition M_i, $i = 1, 2$. Let $M_i = n$ and $M_2 = m$. Let v_{nm} be the value of this game and let $(a_{nm}, 1)$ be the support of the optimal (or -optimal) strategies for the players. Of interest is the asymptotic behavior of the quantities a_{nm} and v_{nm} as m and n go to infinity. This behavior is especially important in the discrete case because of the rapidly increasing difficulty in obtaining explicit solutions to such games as m and n get large.

Without loss of generality, it can be assumed that $A_1(t) = -\log(1-t)$. For non-discrete duels, Kimeldorf and Lang (1977) proved the following result: Let A_2 be a modified accuracy function of the form $A_2(t) = ct^r + h(t)$ where $c > 0$ and $r > 1$ are fixed constants and $h'(t) = o(t^{r-1})$ as $t \to 0+$. Let $M_1 = n$, $M_2 = m_n$, and let $(a_n, 1)$ be the support of the optimal pure strategies and v_n be the value of the game. If $n \to \infty$, $m_n \to \infty$, and $v_n \to v$ where $-1 < v < 1$, then

$$\lim_{n\to\infty} na_n = r\delta\alpha \tag{2.14}$$

and

$$\lim_{n\to\infty} m_n n^{-r} = c^{-1}r^{-r}\delta^{1-r} \tag{2.15}$$

where

$$\alpha = \alpha(v,r) = [(1-v)/(1+v)]^{1/(r-1)}$$

and

$$\delta = \delta(v,r) = \int_{\alpha(v,r)}^{\infty} u^{-2}(1 + u^{r-1})^{-1} du$$

This asymptotic result was then extended to the classical noisy, silent and silent-noisy (discrete-fire) duels in Kimeldorf and Lang (1978). Thus, it was shown that for large numbers of bullets the presence or absence of noise is asymptotically irrelevant.

In the late 1970's other authors extended the classical discrete duel models in other directions. Teraoka (1975, 1976a) solved a silent duel in which each player has at most one (but possibly no) bullet. More precisely, the numbers m and n of bullets possessed by the players were assumed to have a bivariate Bernoulli distribution, the parameters of which are known to both. Thus, not only does a player not know whether or not his opponent has a bullet, he does not know whether he himself has a bullet. Teraoka also solved the corresponding duel when the bullets are both noisy (1976) and when one is noisy and the other silent (1981). Teraoka (1979a) solved the duel in which each player had exactly one bullet, but with uncertainty regarding whether the bullet is silent or noisy. Letting $\theta_i = 1$ if Player i's bullet is silent and 0 otherwise, Teraoka assumed that the random variables θ_1, θ_2 have a bivariate Bernoulli distribution, the parameters of which are known to both (so that Player 1 has as much information about his bullet as has Player 2). In his doctoral dissertation, Teraoka (1979b) presents all of the foregoing results in addition to some other related material and an extensive bibliography.

Fox (1980) extended the classical duel models by considering different payoffs. He retained the assumption that the games be zero-sum, but assumed that the payoff to Player 1 is a if he is the sole survivor, $-b$ if Player 2 is the sole survivor, $a-b$ if both are killed simultaneously, and 0 if both survive. He considered the silent duel, noisy duel, a silent-noisy duel with one bullet for the noisy player, and the nondiscrete-fire duel.

Because of the importance of the duel models to problems in government, industry, and military science, psychologists have recently studied the behavior of experimental subjects in simulated conflict environments in which one of the duel models is applicable. The experimenters compared the subjects' behavior with that prescribed by the mathematical solutions. Four students in this area are Rapoport, Kahan, and Stein (1973), (1976), and Kahan and Rapoport (1974), (1975). The latter authors observe that "players wait longer then the minimax prescriptions [of Fox and Kimeldorf (1970)] before firing their bullets." Kahan and Rapoport conclude that "an analysis of conflict situations using the...concepts of games of timing can lead to a better understanding of such situations, and to a formulation of superior strategies when one is in such a conflict situation."

3. OPEN PROBLEMS

3.1. Duels with Both Silent and Noisy Bullets

Except for very special cases, duels in which some bullets are noisy and some silent remain unsolved. The solution to the duel in which Player 1 has only one noisy bullet and Player 2 has only one silent bullet is well known and is described in Blackwell and Girshick (1954), Karlin (1959) and Dresher (1961). Smith (1967) considered duels in which Player 1 has two bullets of which one is silent and the other noisy, while Player 2 has only a noisy bullet. He solved the duels under the restriction that Player 1 must fire his noisy bullet first.

The class of silent m versus noisy n duels, where Player 1 has m silent bullets and Player 2 has n noisy bullets, can be considered intermediate between the m versus n duels with all bullets silent solved by Restrepo (1957) and the m versus n duels with all bullets noisy which were solved by Fox and Kimeldorf (1969). The special case in which $m = 2$ and $n = 1$ was solved by Mihailova

(1973). Styszynski (1974) then generalized Mihailova's result to arbitrary m, still retaining the assumption that $n = 1$. No results are presently available when n is greater than 1; even the case where $n = 2$, $m = 1$, and $P_1(t) = P_2(t) = t$ remains unsolved.

The solution of these duels is more difficult than either the m noisy versus n noisy duels or the m silent versus n silent duels. The noisy duels could be solved recursively because they are almost games of perfect information. The silent duels are tractable because the space of admissible pure strategies for each player is a Cartesian product of intervals. In the mixed duels, the space of admissible pure strategies for Player 1 is much more complex. For example, in the duel in which Player 1 has 1 silent bullet and Player 2 has 2 noisy bullets, an admissible pure strategy for Player 1 is determined by a real number x and a function f of one variable. The number x can be interpreted as the time Player 1 plans to fire if he has not yet been fired upon, and $f(s)$ as the time Player 1 plans to fire if he still has a bullet when fired upon at time s. (Clearly, he will wait to fire at time $t = 1$ if he still has a bullet when fired upon for the second time.)

3.2. Random Numbers of Bullets

The only duels solved to date in which the amounts of ammunition available to the players are unknown are those of Teraoka in which each player has at most one bullet. More generally one could assume instead that the amount of ammunition each player has is unknown to his opponent, and must be considered by him to be a random variable with a certain (prior) probability distribution f. This model would be applicable in conflict situations where a player does not have perfect information about the size of the opponent's arsenal, but only partial knowledge. The distribution f is revised during the course of the duel in accordance with Bayes' Theorem. It would be interesting to find specific parametric classes of distributions for which the values of the game as well as the optimal strategies can be found explicitly. In cases where explicit solutions are unattainable, various qualitative properties of the optimal or ε-optimal strategies could be studied. An interesting feature of these duels is the manner in which f_t varies with t, where f_t is the conditional distribution of the number of bullets a player started with, given that the duel has not ended by time t.

3.3. Semi-Noisy Bullets

In these duels Player 1 has m bullets and Player 2 has n bullets. Associated with Player i is a number p_i which represents the probability that each of Player i's bullets will be detected by his opponent. It is assumed that Player i knows whether or not his opponent has detected his shot. These duels contain as special cases the silent duels, the noisy duels, and the mixed duels proposed above. Only for small values of m and n can we hope to achieve an explicit solution to these games. In certain cases, it may be possible to solve the games implicitly in terms of systems of functional equations and to study various qualitative properties of the solution. Computer simulation might also be useful.

3.4. Duels with Unknown Accuracy Functions

All of the duels considered in the literature assume that the accuracy functions are known. In actual practice, however, it is far more common that the characteristics of an opponent's weapons are unknown and hence one cannot predict in advance the probability that an opponent's actions will succeed. Therefore, a more realistic mathematical model for actual conflict situations would assume only partial information about the opponent's accuracy function $P(t)$, this information being revised as the duel proceeds. Thus, in addition to a game-theoretic problem, we have a statistical problem of estimating the opponent's accuracy $P(t)$ for $t > t_0$ based on the course of the duel up to time t_0. Of course, the statistical and game-theoretic problems are not distinct, but are convolved in that a player will want to behave so as to minimize his opponent's knowledge of his accuracy function.

3.5. Games over Function Spaces

A game over function space is a zero-sum two-person game in which a pure strategy for player i is a (vector-valued) Borel-measurable function F_i on some compact metric space T, and the payoff kernel M is of the form

$$M(F_1, F_2) = \int_T Q(F_1, F_2, t)dW(t)$$

with suitable conditions on Q and W. Only very special classes of such games have been solved. For example, Fleming (1954) solved some particular games of this form and related their solution to solutions of certain problems in the calculus of variations. In order to ensure the existence of a solution, the functional Q is usually assumed to satisfy certain continuity conditions. Lang and Kimeldorf (1976) showed that all silent duels, including discrete-fire, continuous-fire, and mixtures of the two, can be written in this form, where $T = [0,1]$ and where Q and W take relatively simple forms, which reflect the structure of the game as a game of timing. Having solved these classes of duels, it is natural for us to explore the extent to which these methods generalize. In particular, it may be possible to identify classes of games of timing over function spaces which can be solved using suitable extensions of the techniques used in Lang and Kimeldorf (1976).

3.6. Relative Cost Effectiveness of Duels

The most obvious practical applications of the theory of duels are in making strategic decisions in conflict situations in which one of the duel models is applicable. Another, and perhaps more important, application is the optimal allocation of resources in preparing for a conflict situation. For example, Fox and Kimeldorf (1968) initially studied the price of inaccuracy in noisy duels by considering the trade-off between accuracy and the number of bullets.

By considering the game-theoretic values of various duels with differing amounts of noise, ammunition and accuracy, we could study the relative weights to be put on increasing the size of the arsenal versus increasing the accuracy in a wide class of duels, including those which have only recently been solved. A deeper quantitative knowledge of the trade-offs between accuracy and quantity of ammunition would be of great value in strategic arms limitations agreements and in determining the necessary size of an arsenal in order to maintain parity with an enemy's arsenal. One could also evaluate the importance of gathering intelligence concerning the size of the opponet's arsenal and of maintaining secrecy concerning one's own. For example, the results of Kimeldorf and Lang (1978) can be used to show that when the amount of ammunition available to each competitor is large or when continuous firing is permitted, then the values of secrecy and intelligence become negligible. It is expected that many of these questions regarding optimal allocation of resources and the values of intelligence and secrecy will remain valid in conflict situations in which the duel models are not strictly valid, but only approximately so.

REFERENCES

[1] Blackwell, D. and Girshick, M. A., Theory of Games and Statistical Decisions (John Wiley & Sons, New York, 1954).

[2] Dresher, M., Games of Strategy (Prentice-Hall, Englewood Cliffs, New Jersey, 1961).

[3] Fleming, W. H., On a class of games over function space and related variational problems, Annals of Mathematics 60 (1954) 578-594.

[4] Fox, M. and Kimeldorf, G., Strategies and values in noisy duels, Proceeding of the 1968 U.S. Army Operations Research Symposium (1968) 27-34.

[5] Fox, M. and Kimeldorf, G., Noisy duels, SIAM Journal on Applied Mathematics 17 (1969) 353-361.

[6] Fox, M. and Kimeldorf, G., Values and shooting times in noisy duels, Journal of the American Statistical Association 65 (1970) 422-430.

[7] Fox, M., Duels with possibly asymmetric goals, Zastosowania Matematyki 17 (1980) 15-25.

[8] Kahan, J. P. and Rapoport, A., Decisions of timing in bipolarized conflict situations with complete information, Acta Psychologica 38 (1974) 183-203.

[9] Kahan, J. P. and Rapoport, A., Decisions of timing in conflict situations of unequal power between opponents, Journal of Conflict Resolution 19 (1975) 250-270.

[10] Karlin, S., Reduction of certain classes of games to integral equations, in: Kuhn, H. W. and Tucker, A. W. (eds.), Contributions to the Theory of Games II (Ann. Math. Studies, Vol. 28), 1953, 125-158.

[11] Karlin, S., Mathematical Methods and Theory in Games, Programming and Economics (Addison-Wesley, Reading, Massachusetts, 1959).

[12] Kimeldorf, G. and Lang, J. P., Asymptotic properties of nondiscrete duels, Journal of Applied Probability 14 (1977) 153-161.

[13] Kimeldorf, G. and Lang, J. P., Asymptotic properties of discrete duels, Journal of Applied Probability 15 (1978) 374-396.

[14] Lang, J. P. and Kimeldorf, G., Duels with continuous firing, Management Science 22 (1975) 470-476.

[15] Lang, J. P. and Kimeldorf, G., Silent duels with nondiscrete firing, SIAM Journal on Applied Mathematics 31 (1976) 99-110.

[16] Mihailova, A. S., A mixed 2 × 1 duel, Advances in Game Theory (Proceedings of the Second All-Union Conference on Game Theory, 1973), 56-59. (Russian)

[17] Orlowsky, K. and Radzik, T., A k-noisy, n-silent vs. one-noisy duel with equal accuracy functions, Zastosowania Matematyki 15 (1976) 55-64.

[18] Rapoport, A., Kahan, J. P., and Stein, W. E., Decisions of timing in conflict situations of incomplete information, Behavioral Science 18 (1973) 272-283.

[19] Rapoport, A., Kahan, J. P., and Stein, W. E., Decisions of timing in experimental probabilistic duels, Journal of Mathematical Psychology 13 (1976) 163-191.

[20] Restrepo, R., Tactical problems involving several actions, in: Contributions to the Theory of Games, III (Ann. Math. Studies 39), 1957, 313-335.

[21] Shiffman, M., Games of timing, in: Kuhn, H. W. and Tucker, A.W. (eds.), Contributions to the Theory of Games II (Ann. Math. Studies, Vol. 28), 1953, 97-123.

[22] Smith, G., A dual with silent-noisy gun versus noisy gun, Colloquium Mathematician 17 (1967) 131-146.

[23] Styszynski, A., An n-silent-vs.-noisy duel with arbitrary accuracy functions, Zastosowania Matematyki 14 (1974) 205-225.

[24] Sweat, C. W., A duel involving false targets, Operations Research 17 (1969) 478-488.

[25] Teraoka, Y., Silent duel with uncertain shot, Rep. Himeji Inst. Tech., 28 A (1975) 1-4.

[26] Teraoka, Y., Correction to silent duel with uncertain shot, Rep. Himeji Inst. Tech. 29 A (1976) 1-4. (a)

[27] Teraoka, Y., Zero-sum games of duel with asymmetric existence of the shot fitted to player's gun, Rep. Himeji. Inst. Tech. 29 A (1976) 5-10. (b)

[28] Teraoka, Y., Noisy duel with uncertain existence of the shot, Internat. J. Game Theory 5 (1976) 239-249. (c)

[29] Teraoka, Y., A single-bullet duel with uncertain information available to the duelists, Bull. of Math. Statist. 18 (1979) 69-83. (a)

[30] Teraoka, Y., Contributions to the mathematical analysis of duels: games of timing, Doctoral dissertation, Himeji Institute of Technology, Japan, 1979. (b)

[31] Teraoka, Y., Silent-noisy duel with uncertain existence of the shot, Bull. of Math. Statist. (1981).

[32] Yanovskaya, Ye. B., Duel-type games with continuous firing, Izvestija Akademii Nauk SSSR., Technicheskaya Kibernetika, No.1, 16-19 (Russian). English translation in Engineering Cybernetics (1969), No. 1, 15-18.

MATHEMATICS OF CONFLICT, Martin Shubik (ed.)
Elsevier Science Publishers B.V. (North-Holland), 1983

REQUIREMENTS FOR THE THEORY OF COMBAT

H. K. Weiss

INTRODUCTION

The purpose of this paper is to discuss some of the elements that should be included in a mathematical theory of land combat. The emphasis in the discussion will be on the "engagement" and not on the war. We shall, therefore, be concerned with the middle ground that lies between the now accepted technical methods for evaluating the single shot kill probabilities of weapons, and the valiant efforts that are made from time to time to predict the outcome of whole wars by mathematical methods.

It is the "engagement" with which Clausewitz is principally concerned in his study "On War"; the philosophy with which he supported his study is as appropriate today as it was in 1830 when it was completed. For example, consider what he has to say about the value to be placed upon theory:

> If theory investigates the things that make up war, if it separates more distinctly that which at first sight seems confused, if it explains fully the properties of the means, if it shows their probable effects, if it clearly defines the nature of the ends in view, if it sheds the light of a deliberate critical observation over the whole field of war--then it has achieved the main object of its task. It then becomes a guide to whoever wishes to become familiar with war from books; it everywhere lights up for him the road, facilitates his progress, educates his judgement, and keeps him from going astray.
>
> If principles and rules result of themselves from the observations that theory institutes, if the truth of itself crystallizes into these firms, then theory will not oppose this natural law of the mind. It will rather, if the arch ends in such a keystone, bring it out more prominently; but it does this only to satisfy the philosophical law of thought, in order to show distinctly the point to which all lines converge, not in order to construct from it an algebraical formula for use upon the battlefield. For even these principles and rules serve more to determine in the reflecting mind the leading outlines of its accustomed movements than, like signposts, to point the way for it to take in execution. (Clausewitz, 1830, p. 77.)

By "theory," therefore, we need not mean "mathematical theory." We deal, however, with the combined effect of many variables. The ability of separate human minds to agree upon the combined effect of items known only qualitatively as imperfect; a mathematical model has the advantage of making specific the values assigned to each quantity considered important.

But whether our theories are mathematical or not, our hope is that they will illuminate the problem, and identify its essential elements, and we judge their success by the degree to which they do this.

However, our purpose in seeking a sound theory differs from that of Clausewitz; we hope to use it as a guide for the development of new weapons. If, incidentally, it turns out that some of our findings illuminate the nature of tactics, we shall be gratified. Our first concern is that the theory make the proper use of tactics to illuminate the effect which weapon characteristics have on the outcome of the battle.

The potential range of variation in weapon characteristics which we have at our disposal is great, but characteristics are not to be chosen independently. We cannot design an artillery piece for range without affecting mobility; we pay for high terminal effect in a guided missile in terms of guidance system complexity and warhead weight.

A theory is valuable to us in proportion to the number of weapon characteristics that it includes. An overriding criterion of value, however, is the extent to which the theory agrees with experience.

We have, for example, moderate confidence in methods for the computation of probabilities of hitting a target with an artillery piece, for we can check the theory at our convenience on the Proving Ground. We must, however, have grave doubts about any theory that claims to predict the outcome of a war, and give it its proper position, as a fixture to organize and support a systematic discussion of the elements which influence the chances of winning a war.

In the middle ground of the "engagement," where tactics and weapon characteristics intermingle, theory can obtain support from experience. Here historical research serves its purpose, in providing examples against which to test a theory. "Operations Research," which records quantitatively the variables of an engagement (preferably during, or immediately subsequent to the engagement) similarly provides numerical data which can be associated with a theory.

Insofar as simple theories check experience recorded in this way, they may be used as guides to extrapolate to new situations, and with new weapons. Any prediction of the future is erroneous to some degree, however, and so all methods must be subject to constant check against new data, and appropriate revision.

DIVISION OF CHARACTERISTICS

The field for study may be considered to have the following divisions: 1. weapon; 2. user; 3. terrain; 4. target; and 5. enemy.

Some of the physical characteristics in each category can be listed independent of the others. For example, the number of weapons, users, targets, and enemy soldiers can be listed. But the effectiveness of a single weapon depends upon its target, upon the terrain, and in a manner difficult to measure, upon the morale of its user and the enemy. In spite of this natural redundancy in the table, it serves as a useful method for listing in more detail the elements which a theory should consider.

Weapon Characteristics

A first division of weapons may be according to kind. Thus:

1. Kind of Weapon
   a. Hand Grenade
   b. Rifle
   c. Machine Gun
   d. Mortar
   e. Howitzer (Field Artillery Rocket)
   f. Gun (Recoilless Rifle)
   g. Tank
   h. Self-propelled Artillery
   i. Anti-tank Gun (Rocket, Recoilless Rifle)
   j. Anti-aircraft Weapons
   k. Mines (Anti-tank, Anti-infantry)

Weapons may also be divided according to operating characteristics, in which case some of the categories listed above represent merely graduations of effect.

2. Weapon Operating Characteristics
   a. Terminal Effect
   b. Range
   c. Accuracy
   d. Rate of Fire
   e. Type of Trajectory
   f. Weight
   g. Cost
   h. Mobility
   i. Ammunition Weight
   j. Weapon Life

Terrain

Terrain is one of the primary characteristics which establishes the nature of land warfare.

3. Terrain-affected Characteristics
   a. Sighting Distances
   b. Cover (Natural or Artificial)
   c. Mobility

Targets

Targets may be listed according to kind, or vulnerability.

4. Kind of Targets
   a. Personnel (in open, in foxholes, in dugouts, in fortifications)
   b. Weapons
   c. Vehicles
   d. Tanks (i.e., armored vehicles)
   e. Supply dumps and supply lines
   f. Communications and Communication Centers
   g. Roads, Railroads, Bridges
   h. Buildings
   i. Airfields
   j. Command Posts

The vulnerability of a particular kind of target may vary rapidly with time. Troops well intrenched in a defensive position represent an entirely different target from troops attacking. Troops newly arrived in an area are more vulnerable than the same troops after they have had time to dig in.

User and Enemy

The characteristics of the men involved on both sides of an engagement are a primary variable. It is the presence of these men which makes any theory that encompasses them subject to grave doubts as to its universal validity. The men change the problem from one that could be predominantly analytic to one that is fundamentally statistical. Yet it is possible to say a good deal about the men and the way in which they affect the outcome of an engagement. Some of their important characteristics are:

5. Characteristics of Men
   a. Morale
   b. Freshness
   c. Number
   d. Communication

These elements will be discussed in the following paragraphs. Here it is instructive to note:

> The military machine, the army, and all belonging to it, is fundamentally simple, and appears on this account easy to manage. But let it be borne in mind that no part of it consists on one piece, that it is all composed of individuals, each of whom maintains his friction in all directions...the battalion always remains composed of a number of men, of whom, if chance so wills, the most insignificant is able to cause delay or some irregularity. (Clausewitz, p. 54.)

Specification of Weapon Effectiveness

The elements listed above describe the components out of which an engagement is built. In order to begin to combine them into an organized theory we need a coordinating principle. It will be simply stated that the object of combat is the overcoming of the opponent, "the destruction of his military force, whether it be by death or wounds or any other means, whether it be completely or only to such a degree that he no longer wishes to continue the combat...the opponent in a particular engagement is the military force which stands opposed to us" (Clausewitz, p. 174).

We shall be concerned with tactics rather than strategy. "Strategy determines the point where, the time when, and the military force with which the battle is to be fought. By this threefold determination it has therefore, a very essential influence on the issue of the engagement" (Clausewitz, p. 137). Variations in place, time and number we shall, however, consider variables in a theory. We shall be concerned with opposing forces which have come into position before each other, their number and the character of their weapons determined and each with the object of the destruction of the opposing force.

A great deal of work has been done on the specification of the effectiveness of single weapons against single targets. We are able, for example, to prepare curves showing the single shot probability of killing a man, as a function of distance, when the weapon used against him is a hand grenade, a rifle, a carbine, or a howitzer.

We can compute the expected number of targets killed in an area, with a howitzer projectile, and the number of projectiles required to kill at least a specified fraction of the number of targets in an area if we know the kind and density of targets.

However, when it comes to the combination of these characteristics with terrain features, and the capabilities of the men who use the weapons, our methods are on more argumentative grounds. But "...it is a miserable philosophy if, according to the old manner, we establish rules and principles regardless of all moral quantities, and then, as soon as these quantities make their appearance, we begin to count the exceptions, which we thereby after a fashion scientifically formulate, that is, make them the rule; or if we resort to an appeal to genius, which is above all rules, whereby we are given to understand that rules were not only made for fools, but also must themselves be really folly" (Clausewitz, p. 125).

In fact, there is no absolute obstacle to the inclusion of both terrain and the frailities of human nature in a theory; the difficulty is that these quantities have a range of variation often far greater than any of the purely technical components of the problem. They can be described statistically, in terms of their expected values, means, standard deviations and distribution functions. We can no longer expect, however, to be able to predict the outcome of a single engagement with any accuracy.

We can, however, hope to predict the average outcome of a large number of engagements. In essence, we can estimate the odds in our favor in each of a number of trials, and if we can contrive always to have the odds on our side, in a sufficient number of instances, the cumulative result should be strongly weighted in our favor.

We may add to our specification of quantities influencing the outcome of a series of engagements those of which the probability of seeing a given distance is typical. It is possible, for example, to construct a curve for the probability of seeing at least a distance "r" for a particular type of terrain. The curve depends on the height of the person doing the seeing, and the size of the object being seen, and will vary with these as well as with the geographical region.

A curve of this sort has been applied to tank warfare (Peterson, 1951) by assuming that it can be used to represent the distance at which tanks will first see each other. If it is considered legitimate to combine it with the probability of killing, once an enemy is seen, it will yield the probability of seeing an enemy and then killing him with the first round (provided he doesn't fire first), and when averaged over all ranges will give the average first round single shot probability of killing an enemy.

Now there is an obvious point for criticism in this use of such sighting curves since these are computed on the basis of random movement of the tank. It is conceivable that a tank commander might, if he wished, knowing the position of enemy tanks, always contrive to engage them at greater or smaller ranges than those obtained by purely random movement of the combatants. There is, however, not yet any indication from combat data that this is possible in a sufficient number of cases to make such a curve a poor average representation.

Similarly, one might imagine that the angular distribution of strikes on a tank might be related to tactics of both offense and defense. There is again no reason yet to believe that it is possible to change this distribution radically by desirable tactics.

There is no doubt that terrain limits tactics. How far one can go in giving to terrain an absolute limitation on certain conditions of an engagement is the subject of many intensive studies still underway.

Mention was made of the question of a tank "firing first." One might take the probability of encounter vs. range from a "sighting curve." Then from a chart such as Figure 1, we can obtain the probability of firing between and t+dt for the first, 2nd and 3rd rounds, etc., and thus for each tank determine the probability of firing first, first and second, etc., sum for all combinations combined with the probability of killing when a round is fired, and obtain the overall probability of success.

Data such as is indicated in Figure 1 does not yet exist. It is hoped that it will be obtained from Project "Stalk," an experiment planned by BRL jointly with OCAFF.

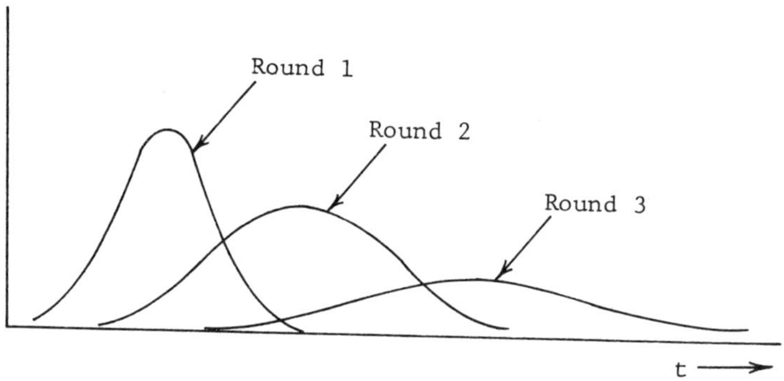

FIGURE 1

However, even when such data is available, it gives only probabilities regarding the outcome of combat between two tanks. The ultimate objective of the theory is the prediction of the characteristics of the complete engagement between large numbers of tanks.

LANCHESTER MODELS

Probably the first attempt to establish a mathematical model of mass combat was made by Lanchester (1916). The model has the following characteristics: Two forces are opposed to each other, consisting of N_r, N_b units respectively. Each unit can fire at any enemy unit, with single shot probability p_r, p_b and rate of fire γ_r, γ_b; separate enemy units can be distinguished, so that this may be called the "aimed fire" case.

Attrition rates on both sides are--

Aimed Fire

$$dn_b/dt = -\gamma_r p_r n_r \quad (1)$$

$$dn_r/dt = -\gamma_b p_b n_b \quad (2)$$

change to fractional attrition

$$d \text{ Ln } n_b/dt = -\gamma_r p_r n_r/n_b \quad (3)$$

$$d \text{ Ln } r_r/dt = -\gamma_b p_b n_b/n_r \quad (4)$$

If these are equal, engagement is a draw

$$\gamma_r p_r n_r^2 = \gamma_b p_b n_b^2 \quad (5)$$

and so numbers are valued as their square, but kill probability and rate of fire as their first powers.

Also write

$$r_b p_b d(n_b^2)/dt = r_r p_r d(n_r^2)/dt \quad (6)$$

$$(N_b^2 - n_b^2)/(N_r^2 - n_r^2) = (\gamma_r p_r)/(\gamma_b p_b) \quad (7)$$

The rate of fire used must be chosen with discretion, since the essence of the method is that when one unit has killed another, it directs its aim at a new target. A mean rate of fire based on average number of rounds per kill and transfer time may be used, or the equations could be complicated to introduce probabilities of kills after various numbers of rounds.

Solving for the number of survivors as a function of time, letting $k = \gamma p$

$$n_b k_b^{.5} = N_b k_b^{.5} \cosh(k_r k_b)^{.5t} + N_r k_r^{.5} \sinh(k_r k_b)^{.5t} \quad (8)$$

and a complementary equation exists for " n_r ."

The time required for one side to be wiped out is given by

$$\tanh(k_b k_r)^{.5t} = (k_b/k_r)^{.5}(N_b/N_r) \quad (9)$$

from which time is seen to depend only on ratio of initial strengths and not on magnitudes.

This is a point subject to check against experience; namely, whether the duration of battles complying with the original specification of the model does indeed tend to be independent of the size of the forces involved, provided the ratios are the same.

From these simple equations one also deduces that if two forces of 1000 dispose of 500, and the survivors then go on to the remaining 500, the side that was divided will lose all of his men while the winner will have lost only 300 men.

This is consistent with the so-called "principle of concentration," stated by Clausewitz (p. 148) as--"The best strategy is always to be very strong, first of all generally, then at the decisive point. Therefore...there is no more imperative and no simpler law for strategy than to keep the forces concentrated."

The above equations hold only when aimed fire can be directed against individual targets. When fire is aimed at an area, then the chance of a target being hit is proportional to the number of targets in an area, as well as the number of firing units, and the Lanchester Equations become

Area Fire

$$dn_b/dt = -c\gamma_r n_r n_b \tag{10}$$

$$dn_r/dt = -c\gamma_b n_r n_b \tag{11}$$

$$d\, \mathrm{Ln}\, n_b/dt = -c\gamma_r n_r \tag{12}$$

$$d\, \mathrm{Ln}\, n_r/dt = -c\gamma_b n_b \tag{13}$$

$$\text{Equating:} \quad \gamma_r n_r = \gamma_b n_b \tag{14}$$

for this case rate of fire and numbers have only equal weight.

FRESHNESS

It is interesting that this development was anticipated by Clausewitz who, however, added the idea of "freshness" of troops. His argument follows:

> If, in an engagement with firearms, 1000 men are used against 500,...1000 men fire twice as many shots as 500, but more shots will strike the 1000...because it is assumed that they stand in closer order...if we were to suppose the number of hits to be double, then the losses on each side would be equal. From the 500...200 disabled, and out of the 1000 the same; now if the 500 had kept another body of equal number in reserve entirely out of fire, then both sides would have 800 effective men; but of those on the one side there would be 500 men quite fresh, fully supplied with ammunition, and in their full vigor; on the other side only 800, all equally disorganized, in want of sufficient ammunition and weakened in physical force. The assumption that the 1000 men, merely on account of their greater number would lose twice as many as 500 would have lost in their place is of course not correct; therefore, the greater loss which the side suffers that has placed half of its force in reserve must be regarded as a disadvantage. Further...in most cases, the 1000 could in the first moment gain the advantage of being able to drive their opponent out of his position and force him to withdraw...there will scarcely be an officer with some experience in war who will not in most cases assign the advantage to that side which has the fresh troops.

> ...this...goes only as far as the disorder, the stage of disintegration and weakness goes, in a word, up to the crisis which every engagement brings with it even for the victor...but where this disintegrating effect of victory coases, and therefore only the moral superiority remains which every victory gives, it is no longer possible for fresh troops to repair the losses;...a beaten army cannot be led back to victory a day after by means of a strong reserve. (p. 150)
>
> The law...is, therefore, that all forces which are available and destined for a strategic object should be <u>simultaneously</u> applied to it; and this application will be so much the more complete the more everything is compressed into one act and into one moment. (p. 154)

With regard to freshness of troops, Marshall (1947) states that 5000 fresh men will defeat 15,000 worn out ones.

Deficiencies in the Lanchester Formula

While we should, perhaps be more pleased that such simple formulae yield reasonable results than critical because of the elements omitted from them, we must look beyond the Lanchester expressions to see how they differ from reality, and what may be added to them.

Some of the deficiencies are listed below:

1. They assume that each unit is within range of all enemy units, and that kill probability does not depend on range.
2. They make no provision for movement, for retreat, or advance.
3. They assume that all units involved on each side are identical, an actual engagement may involve artillery, infantry, armor, minefields, smoke, tactical air, etc.
4. Engagements that continue until one side is wiped out are rare. Retreat begins when the number of casualties approaches the order of 10%.
5. The equations do not give the actual odds (i.e., probabilities) of winning as a function of the variables.

Superiority in numbers does not guarantee victory, but it raises the probability of winning.

> If we strip the engagement of all modifications which it may undergo, according to its purpose, and the circumstances from which it proceeds, if we finally set aside the value of the troops, because that is a given quantity, there remains only the bare conception of the engagement, that is, a combat without form, in which we distinguish nothing but the number of the combatants.
>
> This number will therefore determine victory. Now from the number of abstractions we have had to make in order to arrive at this point, it follows that superiority in numbers is only one of the factors which product victory; that therefore far from having obtained everything or even the principal thing with superiority in numbers, we have perhaps obtained very little by it, according as

> the circumstances involve happen to vary.
>
> But this superiority has degrees, it may be imagined as twofold, threefold, or fourfold, and everyone realizes that by increasing in this way it must overpower everything else.
>
> In this connection we grant that superiority in numbers is the most important factor in the result of an engagement, only it must be sufficiently great to counterbalance all other concurrent circumstances. The direct consequence of this is that the greatest possible number of troops should be brought into action at the decisive point of the engagement.
>
> Whether then these troops suffice or not, we have done in this respect all that our means permitted. This is the first principle in strategy. (Clausewitz, p. 138)

The Lanchester equations, can, in fact be converted to difference-differential equations which deal with the probabilities that exactly " n_b, n_r " units survive to time " t " (Morse and Kimball, 1951). The equation is of the form

$$dP(n_b, n_r, t)/dt = [P(n_b+1, n_r, t) - P(n_b, n_r, t)]\gamma_r p_r n_r + [P(n_b, n_r+1, t) - P(n_b, n_r, t)]\gamma_b p_b n_b \qquad (15)$$

and unfortunately the solution is not simple. This sort of problem is readily handled by digital computers, however, which can in a few minutes run through hundreds of engagements and so give the distribution of outcomes.

The other objections listed to the simple equations above are far more important, however, and will change the process and results more importantly if included.

MODEL OF INFANTRY ATTACK ON WELL DEFENDED POSITION

To emphasize some entirely different aspects of a tactical situation, let us outline a model for an infantry attack on a prepared position. We place N_a men in a rectangular wave of depth D, attacking on a front W. N_d defenders fire, each at a rate γ_d, with a single shot probability that varies with range r. We consider only that part of the attack after the attackers have passed through artillery fire, and are too near the defensive line to be fired upon by artillery. We ignore mines and machine guns.

Mathematics is simplified by assuming that each defender fires against the foremost attacker, shifting this fire to the next in line as the foremost falls. As in the Lanchester model we deal with expectations.

The defender is assumed to be so well-emplaced that the attacker must actually place his troops on the front line to kill a defender. We assume however, that as soon as the first attackers reach in defense MLR its strength begins to deteriorate because of hand-to-hand fighting, and in fact, that the attacker wins if the number of attackers exceeds the number of defenders at the instant the first attacker reaches the MLR. The velocity of the attack wage is " v " ; its leading edge is cut back at a rate depending on the kill rate inflicted on the attacker.

$$dN_a/dt = -N_d\gamma_d p_d(r) \tag{16}$$

so $$dr/dt = -v + N_d\gamma_d p_d(r)D/N_a = -v + \alpha v p(r) \tag{17}$$

where

$$\alpha = \frac{N_d\gamma_d D}{N_a V} \tag{18}$$

If $p(0) = 1$, attack cannot possibly reach MLR unless

$$\alpha < 1.0 \tag{19}$$

physically α = kills/sec at $r = 0$ by defenders divided by men/sec passing a given line parallel to attack front.

Now compute number of survivors in attack at instant front of wave touches MLR. Assume defense only effective to range " r_m " and $p = 1 - (r/r_m)$. Start attack with leading edge at " r_m ," trailing edge at $r_m + D$. Let t_1 be time leading edge reaches $r = 0$, in this time trailing edge moves to $r_m + D - vt_1$

$$T'x + x = 1 - \alpha^{-1} \;;\quad T = r_m/(\alpha v) \;;\quad x = r/r_m \tag{20}$$

$$x = 1 - [(1 - e^{-t/T})/\alpha] \tag{21}$$

whence $$vt_1 = -(r_m/\alpha)\mathrm{Ln}(1-\alpha) \tag{22}$$

for small α ,

$$vt_1/r_m = 1 + (\alpha/2) + (\alpha^2/3) + \ldots \tag{23}$$

The depth of the attack at t_1 is $r_m + D - vt_1$ and the number of survivors is

$$N_a(r_m + D - vt_1)/D \tag{24}$$

for the attacker to win, this must be greater than N_d

$$N_d/N_a < 1 + (r_m/D)[1 + \alpha^{-1}\mathrm{Ln}(1-\alpha)] \tag{25}$$

Three natural parameters are

$$N_a/N_d;\; D/r_m,\; n^* = \gamma_d r_m/v \tag{26}$$

The last, n^* , is the number of rounds one defender fires while one attacker crosses r_m .

$$N_a/N_d > 1 - n^{*\,-1}\left[1+\alpha^{-1}\mathrm{Ln}(1-\alpha)\right] = 1 - n^* g(\alpha) \qquad (27)$$

where $g(\alpha)$ is given by the following table

α	$-\alpha^{-1}\left[1+\alpha^{-1}\mathrm{Ln}(1-\alpha)\right]$
1	
.9	1.75
.8	1.27
.7	1.04
.6	1.86
.5	0.78
.3	0.63
0	0.50

for small α,

$$N_a/N_d > 1 + n^*\left[1/2 + \alpha/3 + \alpha^2/4 + \alpha^3/5 + \ldots\right] \qquad (28)$$

As an illustration, assume that the defense can shoot to 100 yards, and that each defender can fire 5 aimed rounds in the time an attacker takes to cross r_m. The attack has to outnumber the defense at least 7/2, while if the attack wave is 100 yards deep the ratio rises to 7/1.

If the range of the defensive weapon is doubled, so that twice as many rounds can be fired, the required superiority of the attack increases substantially.

If the number of rounds fired by the defender is reduced, the required superiority of the attacker is reduced.

In this connection, it has been observed by Marshall (1947) that only a fraction of soldiers ever fire their rifles in combat. Note also the following remarks by the Italian Marshal, Count Cadorna, quoted by von Lieb (1943, p. 89): "...the operation of the German 14th Army would not have succeeded...had each man fired a clip of cartridges, each machine gun a belt, each gun a shot."

DEFICIENCIES IN MODEL OF ATTACK

The model described above does indicate the difficulty in overwhelming a defense simply by numbers, and conforms to conditions which existed in trench warfare before the stalemate of World War I was broken by the tank.

With two such defenses facing each other one could say--

> I may feel too weak to attack B, from which it does not follow that B is strong enough for an attack on A... it may so happen that both parties, at one and the same time, not only feel themselves too weak to attack, but also are so in reality. (Clausewitz, p. 163)

However, the model suffers from the following deficiencies:

1. It does not include the effect of artillery, of air, or of armor, in reducing the defensive positions.
2. The defensive line does not have depth. It would be particularly vulnerable to attack by unconventional weapons.

There is some reason to believe that in a temporarily stabilized position the density of men decreases with distance back from the front line so that the percentage farther back than " s " is approximately

$$100e^{-s/s_0}$$

where " s_0 " is for one case examined, about 3.5 miles.

It is possible that the use of weapons with very large lethal areas will bring the two front lines in such a situation very close together again, so that the weapon could not be used against the enemy for fear of insuring one's own troops. Probing and reconnaissance actions might then be continuous, to insure that the enemy has not withdrawn from a position in order to attack it with a special weapon.

A THEORY OF MORALE

One of the most difficult elements to introduce to a problem is the morale of the troops. It is possible, however, to make a beginning at including this factor symbolically.

The following approach has been proposed by Rashevsky (1947), who begins with the notion that "fighting ability" is present to varying degrees in men, that in some men it is naturally present, while in others it can be developed.

We may imagine that we could arrange the men in an outfit in an order of decreasing "natural fighting ability." Actually, we have so little information that we are justified in arbitrarily dividing the men into two groups, "actives," who when left to themselves on the battlefield will fight aggressively, and "passives" who will not fight alone without training.

Passives, however, can be inspired by contact with actives and to a lesser extent by contact with other converts. On the other hand they may have a relapse on contact with unconverted passives. Mathematically one may express these observations as follows:

An outfit has A aggressive fighters, out of N men. At any time A_c passives have been converted to actives; contact with A's converts at a rate a , with A_c at a rate a_c .

$$dA_c/dt = aA + a_cA_c - a_c(N - A - A_o) \tag{29}$$

$$= (a + a_c)A + 2a_cA_c - a_cN \tag{30}$$

since converts meeting passives relapse.

Clearly an outfit will not be activated starting from scratch

$(A_c = 0$, at $t = 0)$ unless

$$A/N \geq a_c/(a + a_c) \; ; \quad A/(N-A) \geq a_c/a \tag{31}$$

i.e. if contact with A is five times as inspiring as with an A_c, you only need one fifth as many A and passives to start the outfit.

When an active is killed it has a negative effect on the outfit, equal to positive if he were there, i.e., $-a(A_o - A)$ where A_o is the number the outfit started with. Killing an A_c has a negative effect less in proportion. Assuming a fully "activated" output entering combat,

$$dA_c/dt = -aA_k + a(A - A_k) - a_c A_{ck} + a_c(A_c - A_{ck}) \tag{32}$$

$A_c = N - A$; let A_c be killed at " f " the rate of A

$$A_{ck}/A_c = fA_k/A$$

set eq. = 0

$$A_k/A = \frac{c_1 + (A/N)(1 - c_1)}{2[c_1 f + (A/N)(1 - c_1 f)]} \tag{34}$$

If $f = 1.0$, $A_k/A = 1/2$

$f = 1/2$, $c_1 = 1/5$, $A/N = 1/6$, $A_k/A = 10/15$

It is conceivable that one may introduce the idea of two classes of troops into the combat equations. Rashevsky has suggested that the Lanchester Equations might be written in the form

$$dA_r/dt = -(A_r/n_r)^{\beta}(\gamma)_b p_b n_b \tag{35}$$

$$dA_b/dt = -(A_b/n_b)^{\beta}(\gamma)_r p_r n_r \tag{36}$$

where the exponent " β " is less than one if "actives" are killed at a greater rate as a result of the fact that they furnish the source of aggression in the battle.

If the theory approximated reality, and if $\beta = 1.0$, then half the "actives" would be killed when half their supporting troops had been killed. At this stage demoralization should set in. Actually, smaller losses are required for demoralization, so if killing the "hard core of resistance" is to be considered a cause, "actives" must be considered to have a higher casualty rate than passives.

One suggested way of checking this is the following. Examine records on soldiers decorated for valor in action, and compare their wounding rate, and killed in action rate with those of their comrades who served in the same actions and were not decorated. The assumption is that men who are decorated are among the most valuable fighters in their outfits (the citations can be checked for reason for award). From this data it may be possible to determine the relative loss rates required for Rashevsky's theory.

The above notes suggest that "morale" can be quantitatively related to: a. example set by companions; b. leadership; and c. training and experience.

However, in order for the interaction among the actives and the passives to bring a whole outfit up to fighting trim, communication is necessary. An example of failure of communication that deserves careful study is the case of the U.S. 65th Infantry Regiment in Korea. This outfit, predominantly Puerto Rican fought respectably for two years with the U.S. 3rd Division in Korea. Many of its men spoke only Spanish, however, 1000 green replacements were put into the regiment and there was a shortage of non-coms who could speak both Spanish and English. Ninety-seven members of the regiment in three separate actions were accused of "misbehavior in action," "failure to obey orders," or "leaving the scene of action without proper authority."

It is also worth noting that many commanders objected to the formation of such groups as the "Rangers" composed of men especially chosen for fighting ability, since they considered these few men the hard core around which the fighting strength of their outfits was built.

Requirements for a Theory of Communication

Of all areas of Army operations, that of communications has had probably the least effort devoted to quantitative measures of adequacy. Recent developments of a quantitative theory of communication by Shannon at Bell Telephone Laboratories suggest a method of approach to a measure of a communication.

The unit of communication is the "bit," or simple "yes-no" choice. It is possible to imagine the whole theatre of operations divided into small squares, each of which is queried as to whether or not it contains a man, a tank, a howitzer, etc. This is the maximum information on position of elements of opposing forces that could be given. A second scanning of the whole area at a slightly later time interval would reveal movement. Because men move only a short distance in an hour (say), tanks a greater distance, and pill boxes not at all, the complete information in the first scanning need not be supplied for the second. This leads to a measure of required rate of obtaining information.

In practice, such complete information will never be available; the actual information available through photo and electronic reconaissance, observation, patrols, prisoners of war, documents, sound-flash observation devices, agents, etc. can be measured in "bits" and compared with the optimum. The attempt would be to relate the amount of available information to the chances of success.

False information corresponds to "noise" in an electronic circuit; information may be false in part or in whole. A change in information received may represent a change in the situation, or a change in the error. This also is susceptible to treatment by Shannon's methods.

In conducting an operation, each component unit operates on certain rules and instructions which it carries with it; these are subject to control and change to some extent through the communication links. When a link is broken, the unit operates according to the sum of much more limited information. This also appears subject to quantitative treatment.

Obviously, a great deal of work will be required before results of even minor validity can come out of this sort of a study. A theory of combat that omitted communication, however, would be very much like a circuit diagram of a radio receiver that omitted the connections between elements.

CONCLUSION

In this paper, many of the elements which must be present in an adequate theory of combat have been discussed. Some of the simple approaches which have been made in the past to such a theory have been described. The gap between the models described and the variety of combat situations which must be explained by a satisfactory model is great. On the other hand, data on the performance of system elements is becoming available at a rapid rate. There is no doubt that it will be possible in the future to combine more of the variables of a combat situation in a combat model than is now done; the techniques for such extension of the models do not appear insuperable. The major difficulty remains the testing of models against actual combat records to insure that they properly represent the tactical situations.

REFERENCES

[1] Lanchester, F. W., Aircraft in Warfare: The Dawn of the Fourth Arm, Constable and Company, Ltd., London, 1916.

[2] Marshall, S. L. A., Men against Fire, William Morrow Publishing Co., New York, 1947.

[3] Morse, Phillip M. and G. E. Kimball, Methods of Operations Research, MIT Press and Wiley, 1951.

[4] Peterson, R. H., The range and angular distribution of AP hits in tanks, BRL-MR 580, December 1951.

[5] Rashevsky, M., Mathematical Theory of Human Relations, Principia Press, Inc., Bloomington, Indiana, 1947.

[6] von Clausewitz, Karl, On War, Modern Library, New York, 1830.

[7] von Lieb, Field Marshall General Ritter, Defense, first translation by Stefan T. Possony and Daniel Vilfroy, Military Service Publishing Co., Harrisburg, Pennsylvania, 1943.

MATHEMATICS OF CONFLICT, Martin Shubik (ed.)

LANCHESTER ATTRITION PROCESSES AND THEATER-LEVEL COMBAT MODELS

Alan F. Karr
The John Hopkins University and
Institute for Defense Analyses
Alexandria, Virginia

0. INTRODUCTION

The purpose of this paper is to present some aspects of the theory and applications of Lanchester attrition models from two particular and, we believe, complementary viewpoints. On the one hand we will be concerned with identifying and articulating physical and mathematical assumptions that underlie and lead to certain stochastic attrition models, and on the other hand, we will consider use of such models as attrition equations in four principal computerized simulations of ground-air combat at the theater level.

In seeking to provide sets of underlying assumptions for attrition models we hope primarily to facilitate qualitative understanding of individual models, to provide one sensible basis for coherent comparisons of different models, and to promote reasoned application of the models. Traditionally, criteria for understanding and comparing models have included such factors as ability to reproduce historical or exercise data, comparability of results to those of established models, and, of course, military judgment. The latter has become preeminent because of a general lack of absolute credibility of all models. Understanding assumptions can, we believe, serve as both an alternative to and a component of military judgment. Since models are used despite the lack of data and despite the lack of credence in their predictive capabilities, it seems essential to take full advantage of any method whereby models may be better understood.

The theater-level combat simulations that are treated in this paper are the CONAF Evaluation Model IV (CEM IV), the IDA Ground-Air Model I (IDAGAM I), the Lulejian-I Theater-Level Combat Model, and the VECTOR-2 Theater-Level Combat Model. These have been chosen because all are reasonable well documented and have received substantial use in studies and analyses.

Our presentation in this paper is rather self-contained, except that no results are proved, and the paper is organized in the following manner. Section 1 is an introductory treatment of the classical Lanchester models of combat in terms of systems of differential equations. We have included only the material necessary to understand the later Sections of the paper, so this Section is not to be regarded as a complete treatment of Lanchester differential equations. Indeed, it is highly selective, but serves the purpose stated above. In Section 2 we discuss stochastic analogues of the differential equation models that are in the form of continuous time Markov processes; the sets of assumptions underlying stochastic analogues of the basic Lanchester models are stated and interpreted. Once again the presentation is selective rather than comprehensive.

Section 3 is, in many ways, the conceptual heart of the paper. In it we describe, in terms of underlying assumptions, exact equations for expected attrition, and approximations to the exact equations, for a number of simplified stochastic models of attrition. The simplifications are that the models are static, with no explicit representation of time, and unilateral. Despite these simplifications (or possibly because of them) these models contain essentially all of the attrition equations used in the four theater-level models named above, and in Section 4 we show that this is so. The discussion in Section 4 presents, in abstracted but accurate form, the equations used in the four models for attrition calculations and indicates how these equations relate to the simplified models described in Section 3.

1. CLASSICAL LANCHESTER THEORY

In keeping with the objectives set forth in the Introduction, we restrict our attention to only a few of the many deterministic Lanchester-type differential equations of combat. Excluded are physical effects such as time dependent coefficients and optimal fire allocations; for introductions and further references to these and other effects the reader is referred to Taylor (1980). We begin with the homogeneous "square" and "linear" laws proposed by Lanchester himself (1956). As general notation and terminology we establish the following. The two sides are called "Blue" and "Red" (a practice dating at least to Lanchester (1956)). Let $t \geq 0$ denote time and let

$b(t)$ = number of Blue combatants surviving at time t

$r(t)$ = number of Red combatants surviving at time t .

1.1. Lanchester Square Law

The classical Lanchester square law postulates that combat is described by the differential equations

$$b'(t) = -k_1 r(t) \tag{1.1a}$$

and

$$r'(t) = -k_2 b(t) , \tag{1.1b}$$

where k_1 and k_2 are positive constants independent of time. That is, each side's losses are proportional to the currently surviving strength of the opposition. Dimensional analysis indicates that k_1 is in units of Blue combatants per Red combatant per time (with k_2 analogous), so that one should interpret k_1 as the rate at which one surviving Red combatant destroys Blue combatants. Note that (1.1) postulates that this rate depends neither on time nor on the number of Blue combatants available as targets:

$$-\frac{b'(t)}{r(t)} = \text{rate (at time } t\text{) of destruction of Blue combatants per Red combatant}$$

$$= k_1$$

for all t (unless $b(t) = 0$ or $r(t) = 0$, as discussed below).

Lanchester's rationale for the form of (1.1) is the following:

> The Conditions of Ancient and Modern Warfare Contrasted. There is an important difference between the methods of defence of primitive times and those of the present day which may be used to illustrate the point at issue. In olden times, when weapon directly answered weapon, the act of defence was positive and direct, the blow of sword or battleaxe was parried by sword and shield; under modern conditions gun answers gun, the defence from rifle-fire is rifle-fire, and the defence from artillery, artillery. But the defence of modern arms is indirect: tersely, the enemy is prevented from killing you by your killing him first, and the fighting is essentially collective. As a consequence of this difference, the importance of concentration in history has been by no means a constant quantity. Under the old conditions it was not possible by any strategic plan or tactical manoeuvre to bring other than approximately equal numbers of men into the actual fighting line; one man would ordinarily find himself opposed to one man. Even were a general to concentrate twice the number of men on any given portion of the field to that of the enemy, the number of men actually wielding their weapons at any given instant (so long as the fighting line was unbroken), was roughly speaking, the same on both sides. Under present-day conditions all this is changed. With modern long-range weapons --fire-arms, in brief--the concentration of superior numbers gives an immediate superiority in the active combatant ranks, and the numerically inferior force finds itself under a far heavier fire, man for man, than it is able to return.

Lanchester goes on to state (somewhat circularly) that

> If ... we assume equal individual fighting value, and the combatants otherwise (as to "cover," etc.) on terms of equality, each man will in a given time score, on an average, a certain number of hits that are effective; consequently, the number of men knocked out per unit time will be directly proportional to the numerical strength of the opposing force.

Various attempts have been made (see Taylor (1980), e.g.) to derive (1.1) from more primitive assumptions concerning combat, normally in terms of "point fire" or "area fire" hypotheses. In the judgment of the author, all of these attempts are flawed and are either not rigorous ("area fire" is not a well-defined mathematical concept) or circular (the "assumptions" represent only a verbal restatement of (1.1)). The essential difficulty, it seems, is that combat deals with discrete units, whereas (1.1) allows fractional attrition. One way to avoid this difficulty is to introduce stochastic versions of the sort discussed in Sections 2 and 3 below.

It is possible, nonetheless, to give various interpretations of (1.1) that are consistent with it but are not viewed as assumptions leading to it (see also Bracken (1979) and Taylor (1980)). Here are two such interpretations.

Point Fire/Single Shot Interpretation. Targets must be destroyed individually but either targets are sufficiently numerous or the ability to locate them is sufficiently good that each attacker (while surviving) locates targets at a constant rate, so that (1.1) holds.

Area Fire/Multiple Shot Interpretation. Targets may be destroyed more than one at a time, but are dispersed over some region. However, that area shrinks as the number of targets decreases and is always a constant multiple of the number of surviving targets. Each attacking weapon destroys all targets in a certain area per unit time.

That the first interpretation is consistent with (1.1) is clear. For the second, if a is the lethal area produced per Red combatant per unit of time and if A is the ratio of occupied area to Blue combatants, then assuming coordination of fire,

$$b'(t) = -\frac{(\text{lethal area produced by Red})(\text{Blue strength at } t)}{(\text{area occupied by Blue at } t)}$$

$$= -\frac{ar(t)b(t)}{Ab(t)}$$

$$= -\frac{a}{A}r(t) ,$$

which is the same as (1.1). Note that both interpretations represent an ability to concentrate attacking forces that is consonant with Lanchester's reasoning.

We stress once more that these are interpretations and not sets of underlying assumptions.

Some mathematical consequences of (1.1) will now be given. Because of its particular structure, the system (1.1) is easy to solve in closed form. The solution is:

(1.2a) $$b(t) = b(0)\cosh \lambda t - \alpha r(0)\sinh \lambda t$$

(1.2b) $$r(t) = r(0)\cosh \lambda t - \alpha^{-1}b(0)\sinh \lambda t ,$$

where $\lambda = (k_1k_2)^{1/2}$ and $\alpha = (k_1/k_2)^{1/2}$. Both the physical interpretations of k_1 and k_2 and the form of the solutions (1.2) dictate that one should interpret λ as a measure of the overall intensity of the combat and α as the relative per combatant effectiveness (one Red combatant relative to one Blue combatant). The equations (1.2) make sense only for t less than or equal to the termination time τ given by

$$\tau = \inf\{t : b(t) = 0 \text{ or } r(t) = 0\} ,$$

when one side or the other is annihilated and the combat ceases. It can be shown that

(1.3) $$\tau = \frac{1}{2\lambda}\log\frac{k_2^{1/2}b(0) + k_1^{1/2}r(0)}{|k_2^{1/2}b(0) - k_1^{1/2}r(0)|} ,$$

which we take to be infinite if the denominator is zero.

From (1.1) it follows easily that

(1.4) $$k_2b(t)^2 - k_1r(t)^2 = k_2b(0)^2 - k_1r(0)^2$$

for all t; it is from this relationship (i.e., the fact that $k_2b^2 - k_1r^2$ is an invariant of the system (1.1)) that the name "square law" derives. As consequences of (1.3) and (1.4) one has the following observations:

1) If $k_1r(0)^2 > k_2b(0)^2$, then $\tau < \infty$, $b(\tau) = 0$ and $r(\tau) = [r(0)^2 - \alpha^{-2}b(0)^2]^{1/2} > 0$ (i.e., Red wins a fight to the finish).

2) If $k_1r(0)^2 = k_2b(0)^2$, then $\tau = \infty$ and $b(t) \to 0$, $r(t) \to 0$ as $t \to \infty$ (the combat leads to a drawn state of mutual annihilation).

3) If $k_1r(0)^2 = k_2b(0)^2$, then $\tau < \infty$, $b(\tau) = [b(0)^2 - \alpha^2 r(0)^2]^{1/2} > 0$ and $r(\tau) = 0$ (Blue wins).

Some additional mathematical relationships arising from (1.1) will be presented following a discussion of the linear law, for which the corresponding relationships will also be presented.

1.2. Lanchester Linear Law

By contrast with (1.1), the classical <u>Lanchester linear law</u> is given by

(1.5a) $$b'(t) = -c_1b(t)r(t)$$

and

(1.5b) $$r'(t) = -c_2r(t)b(t),$$

where c_1 and c_2 are positive constants. The units of c_1 are not the same as those of the constant k_1 appearing in (1.1a); c_1 is in units of "per Red combatant per unit time," i.e., the numerator is dimensionless. One way to interpret this is to rewrite (1.5a) as

$$\frac{b'(t)}{b(t)} = -c_1r(t),$$

which shows that c_1 can be thought of as fractional (or percent) Blue attrition per Red combatant per unit time. From (1.5a) we may also observe that

$$-\frac{b'(t)}{r(t)} = \text{rate of destruction of Blue combatants per Red combatant (at time } t\text{)}$$

$$= c_1b(t),$$

which is directly proportional to surviving Blue strength, not independent of it as was the case for the square law.

Lanchester (1956) envisioned the linear law (1.5) as applicable to combat situations in which concentration of attacking forces (of the kind implicit in the previously given interpretations of the square law) is impossible. In his own words:

> The Hypothesis Varied. Apart from its connection with the main subject, the present line of treatment has a certain fascination, and leads to results which, though probably correct, are in some degree unexpected. If we modify the initial hypothesis to harmonise with the conditions of long-range fire, and assume the fire concentrated on a certain area known to be held by the enemy, and take this area to be independent of the numerical value of the forces, then [(1.5) ensues and] the rate of loss is independent of the numbers engaged, and is directly as the efficiency of the weapons. Under these conditions the fighting strength of the forces is directly proportional to their numerical strength; there is no direct value in concentration, qua concentration, and the advantage of rapid fire is relatively great. Thus in effect the conditions approximate more closely to those of ancient warfare.

It is possible to give the following interpretations (not sets of underlying assumptions!) for the linear law.

Point Fire/Single Shot Interpretation. Targets must be attacked individually but are either sufficiently few or sufficiently difficult to locate and attack that each attacking weapon engages targets at a rate proportional to the number of targets present, so that (1.5) holds.

Area Fire/Multiple Shot Interpretation. Targets may be destroyed more than one at a time but are dispersed over some region that does not vary over time, although targets redisperse between shots. Each attacking weapon generates lethal area, all targets in which are destroyed, at a constant rate per unit of time.

To justify the latter interpretation, let a be the lethal area produced per Red combatant per unit of time and let D be the (constant!) area of the region over which Blue combatants are dispersed. Then (again assuming coordination of fire)

$$b'(t) = -\frac{(\text{lethal area produced by Red})(\text{Blue strength at } t)}{(\text{area occupied by Blue at } t)}$$

$$= -\frac{a}{D}r(t)b(t) \ ,$$

which is of the form (1.5). Note that both interpretations do represent inability of the attacking side to concentrate its fire.

We proceed to a mathematical discussion of (1.5). Despite the unpleasantly nonlinear form of (1.5), it can be shown that when $c_1r(0) \neq c_2b(0)$ the solution is given by

$$b(t) = \left[b(0)^{-1}e^{\gamma t} + c_2\gamma^{-1}(e^{\gamma t} - 1)\right]^{-1} \tag{1.6a}$$

and by

$$r(t) = \left[r(0)^{-1}e^{-\gamma t} - c_1\gamma^{-1}(e^{-\gamma t} - 1)\right]^{-1} \tag{1.6b}$$

where $\gamma = c_1r(0) - c_2b(0)$ is a measure of the initial discrepancy between the two forces. When $c_1r(0) = c_2b(0)$ the solution to (1.5) is given by

$$b(t) = \frac{b(0)}{1 + c_2 b(0) t} \tag{1.6c}$$

and by

$$r(t) = \frac{r(0)}{1 + c_1 r(0) t} . \tag{1.6d}$$

Although these solutions do not seem to be as well known as the solution (1.2) to the square law equations, they are not more complicated.

It may be shown from (1.6), but is more easily shown from (1.5), that

$$c_2 b(t) - c_1 r(t) = c_2 b(0) - c_1 r(0) \tag{1.7}$$

for all t, on which expression the name "linear law" is based.

For linear law combats the termination time τ is always infinite, but the following properties hold:

1) If $c_1 r(0) > c_2 b(0)$, then as $t \to \infty$, $b(t) \to 0$ and $r(t) \to r(0) - (c_2/c_1) b(0)$.

2) If $c_1 r(0) = c_2 b(0)$, then $b(t) \to 0$ and $r(t) \to 0$.

3) If $c_1 r(0) < c_2 b(0)$, then $b(t) \to b(0) - (c_1/c_2) r(0)$ and $r(t) \to 0$.

The interpretations are as for the square law.

The two previously given sets of interpretations provide one basis of comparison of the square and linear laws, which may be used to choose one or the other from some specific modeling application. Note that the dual interpretations for each law refute, quite unambiguously, some early assertions that the square law is applicable only to point fire and the linear law only to area fire. Another way to understand the differences between the two is by means of differences between the sets of assumptions underlying stochastic analogues of these attrition equations. For details, the reader is referred to Sections 2 and 3 below.

Force and Attrition Ratios. In view of the attention given to force ratios and attrition ratios, we conclude our discussion of the homogeneous square and linear laws with comparisons in terms of attrition ratios and derivatives of force ratios. For the former, we have for the square law

$$\frac{b'(t)}{r'(t)} = \frac{k_1}{k_2} \frac{r}{b} , \tag{1.8a}$$

while for the linear law,

$$\frac{b'(t)}{r'(t)} = \frac{c_1}{c_2} . \tag{1.8b}$$

These expressions follow immediately from (1.1) and (1.5), respectively. For the square law, the attrition ratio is proportional to the inverse of the force ratio, so that the superior side continues to increase its superiority, which is consistent with Lanchester's arguments concerning concentration of forces. On the other hand, for the linear law, the attrition ratio remains constant over time. This difference might also be used as the basis for choosing one law to use in modeling a particular form of combat.

The situation for force ratios is more complicated, but not intractable. Using (1.1) and (1.4) one can show after some calculation that for the square law

$$(1.9a) \qquad \left(\frac{b}{r}\right)'(t) = \frac{k_2 b(0)^2 - k_1 r(0)^2}{r(t)^2}$$

(this is the change in the force ratio, rather than the ratio of the changes in the forces). Similarly, (1.5) and (1.7) imply that for the linear law

$$(1.9b) \qquad \left(\frac{b}{r}\right)'(t) = (c_2 b(0) - c_1 r(0))\frac{b(t)}{r(t)} .$$

In particular, for the linear law the force ratio is an exponential function of time. For both cases, if the forces are initially equally strong, that is, if $k_2 b(0)^2 = k_1 r(0)^2$ in the square law case and if $c_2 b(0) = c_1 r(0)$ in the linear law case, the force ratio remains constant over time.

We conclude this Section with a brief discussion of analogues of the square law (1.1) and the linear law (1.5) for combats involving heterogeneous forces. Let

M = number of types of Blue combatants,

N = number of types of Red combatants,

$b_i(t)$ = number of Blue combatants of type i surviving at time t,

$r_j(t)$ = number of Red combatants of type j surviving at time t.

1.3. Heterogeneous Square Laws

By formal analogy with (1.1), one version of the heterogeneous square law asserts that

$$(1.10a) \qquad b_i'(t) = -\sum_{j=1}^{N} k_1(i,j) r_j(t) , \quad i = 1, \ldots, M ,$$

and

$$(1.10b) \qquad r_j'(t) = -\sum_{i=1}^{M} k_2(j,i) b_i(t) , \quad j = 1, \ldots, N ,$$

where k_1 and k_2 are matrices (of dimensions $M \times N$ and $N \times M$, respectively) with nonnegative entries, each of which has at least one positive entry. Not all of the entries need to positive, so that some types of Blue combatants, for example, may be invulnerable to some types of Red combatants. One can interpret $k_1(i,j)$ as the rate at which one Red combatant of type j destroys Blue combatants of type i, which not only is independent of the numbers of various types of Blue combatants available as targets, but also must take into account fire directed at other types of Blue combatants.

One must view (1.10) as based purely on analogy with (1.1), and as lacking clear justification for the linearity in the numbers of attacking weapons. In Section 2 we observe that, using analogous stochastic attrition processes, a somewhat more rigorous and persuasive justification for the functional form (1.10) can be provided. In particular, it will be seen that linearity results from a fundamental assumption of mutual probabilistic independence of various combatants on each side. Thus, (1.10) is a possible analogue of (1.1).

However, analysis in Section 2 on the basis of sets of assumptions underlying analogous stochastic attrition processes indicates that (1.10) is not the proper analogue to (1.1), but rather that (2.15) below is.

The solution to (1.10) is difficult to obtain in a useful form; it has, of course, the matrix representation

$$(1.11)\qquad \begin{bmatrix} b(t) \\ r(t) \end{bmatrix} = e^{tK}\begin{bmatrix} b(0) \\ r(0) \end{bmatrix},$$

where K is the matrix

$$(1.12)\qquad K = \begin{array}{c} \\ M \\ N \end{array}\!\!\begin{array}{c} \begin{array}{cc} M & N \end{array} \\ \left[\begin{array}{c|c} 0 & k_1 \\ \hline k_2 & 0 \end{array}\right] \end{array},$$

but while (1.11)-(1.12) may be feasible for use in numerical calculations, they are not useful for qualitative purposes. Moreover, there does not, in general, exist a simple "state law" analogous to (1.4), although it can be shown that if the symmetry condition

$$(1.13)\qquad \sum_{\ell=1}^{N} k_2(\ell,i)k_1(i,j) = \sum_{m=1}^{M} k_1(m,j)k_2(j,i)$$

is satisfied for all i and j, then

$$(1.14)\qquad \sum_{i,j} (k_2(j,i)b_i(t)^2 - k_1(i,j)r_j(t)^2)$$

$$= \sum_{i,j} (k_2(j,i)b_i(0)^2 - k_1(i,j)r_j(0)^2)$$

for all t. The condition (1.13) holds, for example if $M = N$ and if all entries of both k_1 and k_2 are equal to some fixed constant. In other cases, reduction to a system of two equations is possible (Anderson, 1979).

1.4. Heterogeneous Linear Law

The heterogeneous analogue of the linear law (1.5) is the system

$$(1.15a)\qquad b_i'(t) = -b_i(t)\sum_{j=1}^{N} c_1(i,j)r_j(t)\ , \quad i = 1, \ldots, M\ ,$$

and

$$(1.15b)\qquad r_j'(t) = -r_j(t)\sum_{i=1}^{M} c_2(j,i)b_i(t)\ , \quad j = 1, \ldots, N\ ,$$

where c_1 and c_2 are matrices with nonnegative entries (and at least one positive entry). The system (1.15) seems to be unsolvable in closed form and essentially nothing appears to be known about the qualitative nature of the solution. Of course, one can solve it numerically in specific modeling problems.

This concludes our discussion of the classical deterministic Lanchester models. As previously mentioned, a number of additional effects such as mixed laws, optimal control and time dependent coefficients, have not been treated here. References that the reader may wish to consult include Bonder and Farrell (1970), Deitchman (1962), Dolansky (1964), Helmbold (1966), Morse and Kimball (1951), Springall (1968), Sternberg (1970), Taylor (1974, 1976, 1979), and Weiss (1955).

2. CONTINUOUS TIME MARKOV ATTRITION PROCESSES

In this Section we present some continuous time, discrete state space Markov processes that are analogous to the deterministic Lanchester attrition models discussed in Section 1. We give some emphasis to sets of underlying mathematical assumptions that lead to the stochastic attrition processes in a rigorous manner. Using these assumptions one can understand not only the stochastic processes but also the corresponding deterministic processes.

General Comments. For background concerning Markov processes we refer the reader to Cinlar (1975) or Karlin (1976), or any other standard textbook; our exposition matches that in Cinlar. Let $X = (X_t)_{t\geq 0}$ be a regular continuous time Markov process with countable state space E . Associated with X are:

1) The transition semigroup (P_t) defined by

$$P_t(x,y) = P_x\{X_t = y\}\ ,$$

where P_x denotes probability under the condition that $X_0 = x$. The transition semigroup satisfies the Chapman-Kolmogorov equation:

$$P_{t+s}(x,y) = \sum_{z\in E} P_t(x,z)P_s(z,y)\ .$$

2) The infinitesimal generator (matrix) A defined by

$$(2.1)\qquad A(x,y) = \lim_{h\downarrow 0} \frac{P_h(x,y) - I(x,y)}{h}\ ,$$

where I denotes the identity matrix.

The interpretation of (2.1) is based on the fact that for $x \neq y$ and $t > 0$

(2.2) $$P\{X_{t+h} = y | X_t = x\} = A(x,y)h + o(h)$$

as $h \downarrow 0$, so that $A(x,y)$ may be regarded as the "rate" of transition of the Markov process from the state x to the state y. That (2.2) holds is a consequence of the forward equation

(2.3) $$P'_t = P_t A .$$

The backward equation

(2.4) $$P'_t = AP_t$$

also holds, but in the context of attrition processes is less intuitive and less useful computationally.

Study of Markov attrition processes was begun by R. N. Snow (1948) and has proceeded fitfully since that time. The more important references are the book of Kimball and Morse (1951), the dissertation of Clark (1969), a series of papers (Jennings, 1973a, 1973b; Weale, 1971a, 1971b, 1975, 1976, 1977) by members of the Defence Operational Analysis Establishment in Great Britain and papers of the author (1974a, 1976d). The reader is urged to consult these sources for more detailed information than is given in this Section.

We begin by describing the basis on which a stochastic attrition process is deemed to be analogous to a deterministic differential equation model of combat. First, we introduce some notation. Let $E = N^2$ be the set of integer pairs (i,j) with $i \geq 0$ and $j \geq 0$. A Markov process (B_t, R_t) with state space E can be regarded as a model of attrition in combat between a homogeneous Blue side and a homogeneous Red side provided that the paths $t \to B_t$ and $t \to R_t$ be (with probability one) nonincreasing. That is, only attrition can occur and combatants are counted in integral units. Here

B_t = number of Blue combatants surviving at time t,

R_t = number of Red combatants surviving at time t;

both of these are random variables for each t. By virtue of the interpretation (2.2) of the infinitesimal generator A of the process (B_t, R_t), the process is called analogous to a differential model of combat if the form of the generator A and form of the differential equations are sufficiently similar.

For example, the homogeneous square law process described below has generator A given by

(2.5a) $$A((i,j), (i-1,j)) = k_1 j$$

and

(2.5b) $$A((i,j), (i,j-1)) = k_2 i ,$$

where k_1 and k_2 are positive constants. For all other states (ℓ,m) except (i,j) we have

(2.5c) $$A((i,j),\ (\ell,m)) = 0$$

which, since row sums of a generator matrix are zero, forces

(2.5d) $$A((i,j),\ (i,j)) = -(k_2 i + k_1 j)\ .$$

That is, when the attrition process (B_t, R_t) is in state (i,j) (with i Blue survivors and j Red survivors) it can move only to the state $(i-1,j)$, which corresponds to destruction of one Blue combatant, or to the state $(i,j-1)$, corresponding to destruction of one Red combatant. These transitions, by (2.5), occur at the rates $k_1 j$ and $k_2 i$, respectively, and then, using (2.2), the analogy of (2.5a-b) to (1.1a-b) is evident and compelling.

The author's research described in (Karr, 1974a) was directed at understanding stochastic attrition processes and facilitating their use in combat simulation models by describing sets of probabilistic assumptions that lead by rigorous mathematical reasoning to stochastic attrition processes with particular infinitesimal generators. By comparing assumptions one can understand differences among various processes and more rationally choose a process to use in a given modeling situation. To illustrate this procedure, we now list assumptions for homogeneous square law and linear law processes.

2.1. Homogeneous Square Law Process

This is the process (B_t, R_t) with generator A given by (2.5) and ensues from the following assumptions.

1) All combatants on each side are identical.
2) Times between detections by a surviving Red combatant are independent and identically exponentially distributed with mean r_1^{-1}, regardless of the number of surviving Blue combatants (provided that the latter be nonzero).
3) When a Red combatant detects a Blue combatant an instantaneous attack occurs, in which the Blue combatant is destroyed with probability p_1 and survives with probability $1-p_1$. Total loss of contact then takes place immediately.
4) Blue combatants satisfy assumptions 2) and 3) with respective parameters r_2 and p_2.
5) Conditioned on survival, detection and attack processes of all combatants (on both sides) are mutually independent in the probabilistic sense.

To obtain (2.5) one then takes

$$k_i = r_i p_i\ ,\quad i = 1,\ 2\ .$$

which can be interpreted (consistently with (1.1), incidentally) as an instantaneous rate of kill.

For a completely rigorous derivation of (2.5) from these assumptions and for further probabilistic properties of the process (B_t, R_t), we refer the reader to Karr (1974a).

2.2. Homogeneous Linear Law Process

This is the Markov attrition process (B_t, R_t) with generator A given by

(2.6a) $A((i,j), (i-1,j)) = c_1 ij$,

(2.6b) $A((i,j), (i,j-1)) = c_2 ij$,

(2.6c) $A((i,j), (i,j)) = -(c_1 + c_2) ij$,

and otherwise

(2.6d) $A((i,j), (\ell,m)) = 0$.

The analogy to (1.5) is clear. The process results from the following set of assumptions.

1) All combatants on each side are identical.
2) The time required for a particular Red combatant to detect a particular Blue combatant (given survival of both) is exponentially distributed with mean $\tilde{r}_1^{-1}$. Each particular Red combatant detects different Blue combatants independently of one another.
3) A Red combatant attacks every Blue combatant it detects. The engagement is instantaneous, results in destruction of the Blue combatant with probability p_1 and survival of the Blue combatant with probability $1 - p_1$, and is followed by immediate and complete loss of contact. The outcome of the engagement is stochastically independent of all other aspects of the attrition process. Finally, a Red combatant is never harmed in an engagement that it initiates.
4) Blue combatants satisfy assumptions 2) and 3) with respective parameters $\tilde{r}_2$ and p_2 .
5) The detection and attack processes of all combatants initially present are mutually independent.

Then, as shown in Karr (1974a), the equation (2.6) follows with $c_i = \tilde{r}_i p_i$ $(i = 1,2)$, the interpretation of which is an instantaneous <u>one-to-one</u> rate of destruction.

As shown in more detail in Karr (1974a), the linear law process is fundamentally easier to deal with probabilistically because the embedded Markov chain (Cinlar, 1975) is spatially homogeneous; cf. Karr (1974a, 1977a) for consequences of this observation.

Incidentally, the expressions (2.5) and (2.6) are valid only if $i > 0$ and $j > 0$. If $i = 0$ or $j = 0$, the combat has terminated; mathematically this is represented by making (i,j) in absorbing state with

$$A((i,j),\ (k,\ell)) = 0$$

for <u>all</u> states (k,ℓ) .

By comparing the two sets of assumptions one sees immediately that the fundamental difference between the two processes is that in the square law process the distribution of the time required to detect some opponent does not depend on the number of opponents present, whereas in the linear law process the mean time to detect some opponent is proportional to the inverse of the number of opponents (i.e., the detection rate, in the sense of the parameter (= 1/mean) of an exponential distribution, is proportional to the number of opponents). This distinction is in some sense that made by Lanchester, but is now stated in a manner that is more meaningful both physically and mathematically.

Given these sets of assumptions one can use them to decide which process better represents a specific physical combat situation, namely, the one whose assumptions better match the physical reality.

<u>Analysis of Continuous Time Models</u>. Once a stochastic attrition model is defined there are various questions one might pose concerning it. Examples are:

1) For each t , what are the distribution and expectation of B_t and R_t ?
2) Do these expectations satisfy the analogous system of differential equations?
3) What are the variances of B_t and R_t and how large are the fluctuations about the expectations?
4) If we let

$$T = \inf\{t : B_t = 0 \text{ or } R_t = 0\} , \tag{2.7}$$

which is the termination time in a battle fought to annihilation and is analogous to the time τ in (1.3), what are the distribution and expectation of the terminal state $(B_T,\ R_T) = (B_\infty,\ R_\infty)$?

Among the approaches used to answer these questions the following are most prominent.

1) The probabilistic/analytical approach, which attempts to obtain direct answers (Clark, 1969; Goldie, 1977; Karr, 1974a, 1976; Watson, 1976).
2) The approach of numerical integration of the forward equation (2.3) in order to obtain the transition semigroup (P_t) (Jennings, 1973a, 1973b; Weale, 1971a, 1971b, 1975, 1976, 1977; see also Karr, 1977a).
3) Use of Monte Carlo simulations (Clark, 1969; Karr, 1975a, 1975b).
4) Derivation of approximations valid for large numbers of combatants (Goldie, 1977; Karr, 1976a; Watson, 1976).
5) Use of diffusions approximations (Perla, 1977, 1978).

Of these, we will illustrate only the first approach and only one aspect of it. The reader is referred to the listed sources for further details. Our illustration is based on the following very general, but also very useful, result.

Theorem (Karr, 1977a). Let (B_t, R_t) be a Markov attrition process with generator A, let T be the termination time given by (2.7) and let f be a real-valued function defined on the state space of the process. Then

$$\frac{d}{dt}E[f(B_t, R_t)] = E[Af(B_t, R_t); \{T > t\}] . \tag{2.8}$$

Here the E denotes expectation and the notation $E[X;G]$ means $E[X']$, where X' and X agree on the event G and X' is zero elsewhere.

The expression (2.8) is very general and gives a large number of specific and interesting consequences, a few of which we will list and discuss.

If (B_t, R_t) is the square law process with generator A given by (2.5) and if we take $f(i,j) = i$, then (2.8) gives

$$\frac{d}{dt}E[B_t] = -k_1E[R_t; \{T > t\}] , \tag{2.9}$$

an expression first derived by Snow (1948). The equation (2.9) shows that expectations in the square law case do not satisfy the deterministic square law (1.1), but do so approximately for those t small enough that one can neglect the difference

$$E[R_t] - E[R_t; \{T > t\}] = E[R_t; \{T \leq t\}] .$$

By successfully choosing $f(i,j) = i^2$ and $f(i,j) = j^2$ and performing some algebraic manipulations one obtains from (2.8) the expression

$$\frac{d}{dt}E[k_2(B_t^2 + B_t) - k_1(R_t^2 + R_t)] = 0 , \tag{2.10}$$

which is the stochastic analogue the deterministic state equation (1.4).

Similar, but nicer, results obtain for the linear law process (B_t, R_t) with generator A given by (2.6). In particular,

$$\frac{d}{dt}E[B_t] = -c_1E[B_tR_t; \{T > t\}] \tag{2.11}$$

and therefore (by a symmetry argument)

$$\frac{d}{dt}E[c_2B_t - c_1R_t] = 0 . \tag{2.12}$$

Pleasantly, (2.12) is an exact analogue of the linear law state equation (1.7).

Additional results of the sort represented by (2.9)-(2.12) may be found in Karr (1977a).

A further benefit of identifying sets of underlying assumptions is that generalization to heterogeneous processes can be done easily and plausibly by extending the assumptions, rather than purely formally by writing down similar looking equations. As illustrations, we list the assumptions and generators for heterogeneous square law and linear law processes. Let

M = number of types of Blue combatants,

N = number of types of Red combatants,

$B_i(t)$ = number of Blue combatants of type i surviving at time t ,

$B(t) = (B_1(t), \ldots, B_M(t))$,

$R_j(t)$ = number of Red combatants of type j surviving at time t ,

$R(t) = (R_1(t), \ldots, R_N(t))$.

Let $E = N^{M+N}$ be the state space of the process, elements of which will be written as

$$(\ell,m) = (\ell_1, \ldots, \ell_M, m_1, \ldots, m_N) .$$

2.3. Heterogeneous Square Law Process (Traditional Version)

This is the process $(B(t), R(t))$ with generator A given by:

$$A((\ell,m), (\ell_1, \ldots, \ell_i-1, \ldots, \ell_M, m_1, \ldots, m_N)) = \sum_{j=1}^{N} k_1(i,j)m_j \tag{2.13a}$$

for $i = 1, \ldots, M$; also

$$A((\ell,m), (\ell_1, \ldots, \ell_M, m_1, \ldots, m_j-1, \ldots, m_N)) = \sum_{i=1}^{M} k_2(j,i)\ell_i \tag{2.13b}$$

for $j = 1, \ldots, N$. The term (2.13a) corresponds to the transition caused by destruction of one Blue combatant of type i ; (2.13b) represents destruction of one Red combatant of type j . No other transitions from the state (ℓ,m) are possible.

The analogy of (2.13) to (1.10) is quite clear, but what is particularly persuasive is the analogy of the following set of assumptions to the set leading to the homogeneous square law process. In particular, linearity of the terms (2.13) in the numbers of opposing weapons is a consequence of the independence Assumption 6), which shows that this process cannot represent synergistic effects among combatants on the same side. Here are the assumptions.

1) All combatants of a given type on a given side are identical.
2) Consider a Red combatant of type j . The times between detections it makes of Blue combatants of type i are independent and identically exponentially distributed with mean $r_1(i,j)^{-1}$.
3) For each Red combatant these M ongoing detection processes occur simultaneously and independently.

4) Whenever a Red combatant of type j detects a Blue combatant of type i there occurs an instantaneous attack in which the Blue combatant is destroyed with probability $p_1(i,j)$. Loss of contact follows immediately.

5) Blue combatants satisfy Assumptions 2)-4) with parameters $r_2(j,i)$, $p_2(j,i)$.

6) Detection and attack processes of all combatants are mutually independent.

It is shown in Karr (1974a) that these assumptions imply (2.13) with

$$k_1(i,j) = r_1(i,j)p_1(i,j)$$

and

$$k_2(j,i) = r_2(j,i)p_2(j,i) \ .$$

Only a cursory examination of the assumptions reveals that they are <u>unnatural</u>. Assumptions 2) and 3) imply simultaneously occurring detection processes for each type of opposition combatant, independent of the number of opposition combatants of each type, which seems very unreasonable physically in most combat situations. If the total number of detections does not depend on the total number of combatants on the opposing side, then much more plausible would be assumptions to the effect that times between detections (without reference to the type of opponent detected) are merely exponentially distributed with mean $r_1(j)^{-1}$ for type j Red combatants and that when a detection occurs, and if there are ℓ_i currently surviving Blue combatants of type i, then the detected Blue combatant is of type i with probability $\ell_i / \sum_{n=1}^{M} \ell_n$. Making this modification yield the following.

2.4. Heterogeneous Square Law Process with Fire Allocation

Assumptions 1), 4) and 6) hold as above, together with the following.

2') Times between detections by a Red combatant of type j are independent and identically exponentially distributed with mean $r_1(j)^{-1}$, regardless of the numbers of Blue combatants present.

3') If a detection occurs when the surviving Blue force has composition $(\ell_1, \ldots, \ell_M)$ the detected combatant is of type i with probability $\ell_i / \sum_{n=1}^{M} \ell_n$.

5') Blue combatants satisfy assumptions 2'), 3') and 4) with parameters $r_2(i)$ and $p_2(j,i)$.

Subject to these assumptions the attrition process $(B(t), R(t))$ has generator A given by

(2.14a) $$A((\ell,m),\ (\ell_1,\ \ldots,\ \ell_i-1,\ \ldots,\ \ell_M,\ m_1,\ \ldots,\ m_N)) = \frac{\ell_i}{\hat{\ell}} \sum_{j=1}^{N} r_1(j)p_1(i,j)m_j$$

for $i = 1, \ldots, M$ and by

(2.14b) $$A((\ell,m),\ (\ell_1,\ \ldots,\ \ell_M,\ m_1,\ \ldots,\ m_j-1,\ \ldots,\ m_N)) = \frac{m_j}{\hat{m}} \sum_{i=1}^{M} r_2(i)p_2(j,i)\ell_i,$$

for $j = 1, \ldots, N$, where $\hat{\ell} = \sum \ell_i$ and $\hat{m} = \sum m_j$.

By virtue of the assumptions (which are clearly square law assumptions) the factors $\ell_i/\hat{\ell}$ and $m_j/\hat{m}$ in (2.14) do not make this a linear law process. These factors simply represent a <u>proportional allocation of fire</u>.

The plausibility of the assumptions of this process suggests that (1.10) is <u>not</u> the appropriate deterministic square law in the heterogeneous case, but that instead the appropriate system of equations is

(2.15a) $$b_i'(t) = -\frac{b_i(t)}{\hat{b}(t)} \sum_{j=1}^{N} k_1^*(i,j)r_j(t), \quad i = 1, \ldots, M$$

and

(2.15b) $$r_j'(t) = -\frac{r_j(t)}{\hat{r}(t)} \sum_{i=1}^{M} k_2^*(j,i)b_i(t), \quad j = 1, \ldots, N,$$

where k_1^*, k_2^* are matrices with positive entries and where $\hat{b}(t) = \sum b_i(t)$, $\hat{r}(t) = \sum r_j(t)$. That (2.15) seems to be more appropriate than (1.10) would have been difficult to discern without identification of sets of underlying assumptions.

One can replace the proportional fire allocation of Assumption 3') by a more general allocation of the form

$$w_{ji}\ell_i / \sum_{n=1}^{M} w_{jn}\ell_n,$$

where the w_{jn} are weighting factors. Indeed, it is shown in Karr (1974a) that all allocations satisfying some simple and intuitive hypotheses are of this form. For an illustration we refer to the discussion of the IDAGAM I model in Section 4 below.

We next consider a corresponding linear law model.

2.5. Heterogeneous Linear Law Process

This process has generator A given by

(2.16a) $$A((\ell,m),\ (\ell_1,\ \ldots,\ \ell_i-1,\ \ldots,\ \ell_M,\ m_1,\ \ldots,\ m_N)) = \ell_i \sum_{j=1}^{N} c_1(i,j)m_j$$

for $i = 1, \ldots, M$, and by

(2.16b) $$A((\ell,m),\ (\ell_1,\ \ldots,\ \ell_M,\ m_1,\ \ldots,\ m_j-1,\ \ldots,\ m_N)) = m_j \sum_{i=1}^{M} c_2(j,i)\ell_i$$

for $j = 1, \ldots, N$, with all other transitions impossible. While the resemblance to (1.15) is clear, the analogy of the following assumptions to the assumptions engendering the homogeneous linear law process is even more compelling.

1) All combatants of a given type on the same side are identical.
2) The time required for a Red combatant of type j to detect a given Blue combatant of type i is exponentially distributed with mean $\tilde{r}_1(i,j)^{-1}$. Each Red combatant detects different Blue combatants independently of one another.
3) A Red combatant attacks every Blue combatant it detects. The attack is instantaneous and destroys a Blue combatant of type i with probability $p_1(i,j)$; immediately thereafter contact is lost.
4) Blue combatants satisfy Assumptions 2) and 3) with parameters $\tilde{r}_2(j,i)$ and $p_2(j,i)$.
5) Detection and attack processes of all combatants are mutually independent.

As proved in Karr (1974a), (2.16) then follows with $c_1(i,j) = \tilde{r}_1(i,j)p_1(i,j)$ and $c_2(j,i) = \tilde{r}_2(j,i)p_2(j,i)$.

In this case the assumptions are quite compatible with those for the homogeneous case and no alternate version is necessary.

The continuous time stochastic attrition processes are sufficiently intractable to make their application in iterative simulations of large scale combat discouragingly difficult. One alternative is to choose a small time interval and to approximate one of the deterministic equations from Section 1. Another alternative is to seek to develop equations that are simpler (in the sense, for example, that they contain no explicit representation of time) but are still based on probabilistic assumptions that are felt to be plausible. Such assumptions might be regarded as retaining the essential features of the sets of assumptions presented in this Section; alternatively, they might be viewed as more plausible *per se* than the assumptions underlying the continuous time processes given above. In the next Section we develop a number of simplified stochastic models of attrition.

3. SIMPLIFIED STOCHASTIC ATTRITION MODELS

In this Section we present a number of simplified, stochastic attrition models that are related--in terms of the sets of underlying assumptions--to the continuous-time Markov attrition models discussed in the preceding Section. The essential simplification in the models of this Section (as compared to those of Section 2) is that there is no explicit representation of time. Rather, these are static attrition models that represent the attrition occurring over an internal of time only by the total attrition at the end of that interval. Furthermore, the models presented here are one-sided, describing the attrition to a set of targets inflicted by a set of searchers. One way to use such models to describe attrition in bilateral combat is to first view the Blue side as targets and the Red side as searchers (which allows calculation of attrition to the Blue side) and then--using the same initial numbers--to view the Blue side as searchers and the Red side as targets. (Other ways include allowing only the survivors on one side to return fire against the other, or using weighted averages of several different methods.) Since essentially all theater-level combat models are fixed time step, iterative simulations, the models described in this Section are well-suited to use in large scale combat simulations and, indeed, are frequently so used, as we will describe in Section 4 below. In such applications, of course, the numerical values of the parameters of the equations must be chosen consistently with the time step of the simulation.

We will now describe several simplified attrition models, each of which is related to one of the stochastic attrition processes discussed in Section 2 and related in two ways to one of the deterministic Lanchester equations given in Section 1. The first relation is through the stochastic process and the second is a direct relation obtained by approximating the functional form of the model by some simpler form. As general notations and terminology, we introduce:

s = number of searchers

t = number of targets

Δt = attrition to targets.

In our discussion, s and t are not random; only the attrition Δt is a random variable.

The models we are about to describe have been developed mainly in Karr (1974b, 1976b, 1977c); see also Anderson <u>et al</u>. (1974) and Jennings (1973a). Several of the approximations noted below, however, have been in widespread use before underlying assumptions were claified. In the discussions of approximations, when a quantity is termed "small" we mean roughly that its square is sufficiently close to zero that it can be neglected.

3.1. Homogeneous Square Law Process

The assumptions underlying this process are the following.

1) All targets are identical; all searchers are identical.
2) Each searcher detects at least one target with probability $d \in (0,1]$, regardless of the number of targets. For a given searcher each target is equally likely to be in the class of detected targets.
3) A searcher making no detection makes no attack.

4) If a searcher makes at least one detection, it chooses randomly from the targets it has detected, exactly one target to attack. Each detected target is equally likely to be chosen.

5) The probability that a searcher destroys a target, given an attack, is $k \in (0,1]$.

6) The searchers are mutually independent in the probabilistic sense.

The reader will note the essential similarity of the assumptions here to those underlying the homogeneous square law process of Section 2: in the latter the distribution of the time required to make a detection is independent of the number of targets, and in the case at hand, the probability of a detection is independent of the number of targets.

From these assumptions it follows that

$$(3.1) \qquad E[\Delta t] = t\left[1 - \left(1 - \frac{kd}{t}\right)^s\right]$$

the derivation, which is not difficult, is given in Karr (1976b).

While the analogy to square law processes is quite clear in the context of underlying assumptions, some mathematical approximations to (3.1) help to make the analogy still stronger. When the quantity kd/t is small the approximation

$$(3.2) \qquad E[\Delta t] \simeq t\left[1 - \exp\left(-\frac{kds}{t}\right)\right]$$

is valid. If the quantity kds/t is small (which implies that kd/t is small) then (3.1) may be approximated as

$$(3.3) \qquad E[\Delta t] \simeq kds ,$$

which is formally the same as the classical square law (1.1).

We observe that kd/t is small if the number of targets is large (which is consistent with the underlying assumptions; if there are many targets the probability of detecting at least one is not a function of their exact number) or if the detection probability d is sufficiently small. The latter point is important because d is related to the length of the time period over which attrition is being calculated. For short term periods, d can be very small and in this case (3.2) is a useful approximation. It will be seen throughout this Section that approximations--especially exponential approximations--all improve as the appropriate detection probabilities decrease. Of course, if the probability k of target destruction given attack is small, this might also render (3.2) valid, but k is essentially independent of the time period under consideration and is not necessarily small in realistic situations.

Finally, we point out that validity of (3.2) is independent of the value of s , so that (3.2) holds even for large values of s , whereas (3.3) cannot be valid for large s unless kd/t is extremely small, in which case (3.2) becomes a still better approximation. Therefore, (3.2) should be preferred to (3.3) in essentially all computational situations.

3.2. Homogeneous Linear Law Process

Here are the assumptions that underlie this process

1) All targets are identical and all searchers are identical.
2) The probability that searcher i detects target j is $\tilde{d} \in (0,1]$ for all i and j.
3) A given searcher detects different targets independently of one another.
4) A searcher making no detection is unable to attack any target.
5) If a searcher makes at least one detection then it chooses randomly, from among those targets it has detected, exactly one target to attack.
6) Given that a searcher attacks a target, the probability of destroying it is $k \in (0,1]$.
7) Searchers are mutually independent.

Assumptions 2) and 3) are in the spirit of the one-on-one detection mechanism postulated for the homogeneous linear law Markov attrition process (see 2.2 above). To clairfy the nature of the similarity between the two models, we offer the following comparison. For the continuous time process, the probability that a given searcher, over a time interval of length h, detects none of t currently surviving targets is $e^{-(\tilde{r}h)t}$, where $\tilde{r}$ is the one-on-one detection rate. Therefore, the probability of at least one detection (so that a target can be attacked) is

$$(3.4) \qquad 1 - e^{-(\tilde{r}h)t} \simeq \tilde{r}ht$$

if $\tilde{r}ht$ is small. For the model defined by the preceding set of assumptions, the probability that a searcher fails to detect any of t targets is $(1-\tilde{d})^t$, so the probability of at least one detection (and therefore an attack) is

$$(3.5) \qquad 1 - (1-\tilde{d})^t \simeq \tilde{d}t$$

if $\tilde{d}t$ is small. The resemblance of (3.4) and (3.5) to one another and to the linear law (1.5) serves to confirm that the model at hand is a linear law model. Note also that the expressions make explicit the relation between the detection probability $\tilde{d}$ and the length h of the time period under consideration.

Although the derivation (cf. Karr, 1974b) is more complicated than that of the square law model, one can show that based on the above assumptions,

$$(3.6) \qquad E[\Delta t] = t\left\{1 - \left(1 - \frac{k}{t}[1-(1-d)^t]\right)^s\right\} .$$

Approximations to (3.6) include the following.

1) For large values of t or small values of $\tilde{d}$ the approximation

$$(3.7)\qquad E[\Delta t] = t\left[1 - \exp\left(-\frac{ks}{t}[1 - e^{-\partial t}]\right)\right]$$

holds.

2) If ∂ is sufficiently small that ∂t and $k\partial s$ both are small, then one has the approximation

$$(3.8)\qquad E[\Delta t] \simeq k\partial ts ,$$

which, we hardly need mention, is of precisely the same form as the classical linear law (1.5).

As was true for the approximations to the analogous square law model, both (3.7) and (3.8) are better approximations as ∂ (i.e., the length of the time interval) decreases. However, note that (3.7) is also valid for large values of t, which is compatible with theater-level combat, whereas validity of (3.8) required that both ∂t and $k\partial s$ be small, which is not really compatible with theater-level combat. In the limit as $t \to \infty$ there is also the approximation

$$E[\Delta t] \simeq ks ,$$

which is physically reasonable, but too crude even for most theater-level calculations. Moreover, an approximation of this form is valid for essentially every reasonable attrition equation, which makes it useless for comparing and understanding different equations.

The two models we have just described confirm the essential distinction between square law and linear law processes: for square law processes the rate of target detection (or acquisition or localization, to the point that an attack can be initiated) by a given searcher is not a function of the number of targets present, while for linear law processes the rate of target detection is (at least approximately) directly proportional to the number of targets.

As we have tried to point out in Section 2, we believe that a significant advantage of dealing with attrition models in terms of sets of underlying assumption is that doing so makes generalization and extension of the models--by generalizing and extending the assumptions--natural and reasonable rather than artificial and purely formal (as happens when one simply writes down a similar-looking equation and hopes it is correct). To illustrate the point we next give a multiple shot analogue of the homogeneous linear law process.

3.3. Homogeneous Linear Law Process, Multiple Shot Version

The assumptions are the following.

1) All targets are identical; all searchers are identical.
2) The probability that searcher i detects target j is $\partial \in (0,1]$ for all i and j.
3) A given searcher detects different targets independently of one another.
4) A searcher making no detection attacks no target.
5) If a searcher detects one or more targets, the number of those it attacks is the minimum of the number detected and a prescribed integer m. If more than m targets are detected, all subjects of size m of those detected are equally likely

to be chosen to be attacked.

6) Given that it attacks a target, a searcher destroys it with probability $k \in (0,1]$.

7) Different targets attacked by a given searcher are destroyed independently.

8) Searchers are mutually independent.

The previously discussed homogeneous linear law model corresponds to the case $m = 1$. While it is easy to see on the basis of the underlying assumptions that this model indeed does generalize the preceding model to the multiple shot case in which each searcher can attack at most m of the targets it has detected, the generalization is most difficult to perceive (let alone formulate) solely on the basis of formal comparison of the equation (3.6) and the equation (3.9) below.

Subject to the assumptions above, we have

$$(3.9) \qquad E[\Delta t] = t\left[1 - \left(1 - k\tilde{d}\left\{B(m-1;t-1,\tilde{d}) + \frac{m}{t\tilde{d}}B(t-m-1;t,1-\tilde{d})\right\}\right)^s\right] ,$$

where

$$B(\ell;n,p) = \sum_{j=0}^{\ell}\binom{n}{j}p^j(1-p)^{n-j} ,$$

which is the cumulative distribution function for binomial probabilities with parameters n and p . For a derivation of (3.9) the reader is referred to Karr (1977c); see also Jakobovits (1977).

The reader can verify that if $m = 1$, then (3.9) reduces to (3.6). At the other extreme, if $m = t$ (i.e., a searcher attackes every target it detects) then (3.9) becomes

$$(3.10) \qquad E[\Delta t] = t[1 - (1 - k\tilde{d})^s] .$$

Comparison of the equations above when $d = 1$ or $\tilde{d} = 1$, as appropriate, helps further clarify the differences among them. In this case both the exact equations (3.1) and (3.6) reduce to

$$E[\Delta t] = t\left[1 - \left(1 - \frac{k}{t}\right)^s\right] ,$$

which makes it clear that square and linear law processes different only in the mechanism of target detection. The exponential approximations (3.2) and (3.7) become

$$E[\Delta t] \simeq t\left[1 - e^{-\frac{ks}{t}}\right]$$

and

$$E[\Delta t] \simeq t\left[1 - e^{-\frac{ks}{t}(1-e^{-L})}\right] ,$$

respectively, and these two expressions are essentially equal under the conditions for their validity as approximations. On the other hand, (3.3) becomes

$$E[\Delta t] \simeq ks ,$$

while (3.8) becomes

$$E[\Delta t] \simeq kts ,$$

however, $d = 1$ or $\mathfrak{d} = 1$ is generally incompatible with conditions for validity of (3.3) or (3.8). Finally, (3.9) becomes

$$E[\Delta t] \simeq t\left[1 - \left(1 - \frac{km}{t}\right)^{s}\right] .$$

We conclude this Section by discussing heterogeneous versions of the square law and linear law processes.

Let

M = number of types of searchers,
$s(i)$ = number of searchers of type i ,
N = number of types of targets,
$t(j)$ = initial number of targets of type j ,
$\Delta t(j)$ = attrition to targets of type j .

Also, let

$$t = \sum_{j=1}^{N} t(j)$$

be the total initial number of targets. Recall that only the attritions $\Delta t(j)$ are random variables.

3.4. Heterogeneous Square Law Process

The assumptions in this case are essentially the same as those of the homogeneous square law process described earlier in this Section. For searchers of type i , the probability of detecting one or more targets is $d_i \in (0,1]$ and does not depend on the numbers of targets present. Given that a searcher of type i attacks a target of type j , the target is destroyed with probability k_{ji} . We emphasize that the assumption that all searchers be mutually probabilistically independent remains in force. It then follows that for each j

$$(3.11) \qquad E[\Delta t(j)] = t(j)\left[1 - \prod_{i=1}^{M}\left(1 - \frac{k_{ji}d_i}{t}\right)^{s(i)}\right] ;$$

the reader is referred to Karr (1974a) for a derivation.

If each of the quantities $k_{ji}d_i/t$ is small (which certainly occurs in theater-level calculations over relatively short time intervals) then (3.11) admits the exponential approximation

$$(3.12) \qquad E[\Delta t(j)] \simeq t(j)\left[1 - \exp\left(- \frac{1}{t}\sum_{i=1}^{M} k_{ji}d_i s(i)\right)\right] ,$$

which will figure in the discussion in Section 4 of the attrition equations used in several theater-level combat models. Under the stronger condition that $k_{ji}d_i s(i)/t$ be small for each i, one also has the approximation

$$(3.13) \qquad E[\Delta t(j)] = \frac{t(j)}{t} \sum_{i=1}^{M} k_{ji} d_i s(i)$$

to (3.11). It should not be concluded from the appearance of $t(j)$ as a factor of the right-hand side of (3.13) that this equation corresponds to a linear law process. In fact, the underlying assumptions clearly imply that the factor $t(j)/t$ represents a proportional allocation of fire of the same sort that appears in (2.14) and (2.15). Alternative allocations are discussed in Sections 2 and 4.

3.5. Heterogeneous Linear Law Process

The assumptions underlying this process are those underlying the homogeneous linear law process (3.2 above) with modifications to permit heterogeneous forces. Each searcher can still attack at most one target and the probability that a particular searcher of type i detects a particular target of type j is $\mathfrak{d}_i \in (0,1]$. We emphasize that this probability depends only on the type of searcher. Target dependent detection probabilities in general render the process intractable (cf. Karr, 1974b), although it is possible to handle the case where the searcher i-target j detection probability is either $\mathfrak{d}_i$ or zero. See Karr (1974b) for some other variations on the hypotheses. The probability that a searcher of type i destroys a target of type j, given an attack, is k_{ji} (it is essentially always possible to allow kill probabilities to depend on the type of target). These assumptions imply that for each j

$$(3.14) \qquad E[\Delta t(j)] = t(j)\left[1 - \prod_{i=1}^{M}\left(1 - \frac{k_{ji}}{t}[1-(1-\mathfrak{d}_i)^t]\right)^{s(i)}\right];$$

a derivation may be found in Karr (1974b). The reader will note the close resemblance between (3.14) and (3.6); of course, if $M = 1$, the former reduces to the latter.

Two exponential approximations to (3.14) are in use. The more commonly used is given by

$$(3.15) \qquad E[\Delta t(j)] \simeq t(j)\left[1 - \exp\left(-\sum_{i=1}^{M} k_{ji}\mathfrak{d}_i s(i)\right)\right].$$

Conditions for validity of (3.15) are that $\mathfrak{d}_i t$ be small for each i and that $\mathfrak{d}_i k_{ji}$ be small for each i and j. Clearly the first of these (which requires that $\mathfrak{d}_i$ be extremely small) essentially always implies the second. The less frequently used approximation is valid under wider conditions; it requires only that each $\mathfrak{d}_i$ be small (and improves as t increases as well) and is given by

$$(3.16) \qquad E[\Delta t(j)] \simeq t(j)\left[1 - \exp\left\{-\frac{1}{t}\sum_{i=1}^{M} k_{ji}s(i)(1 - e^{-\mathfrak{d}_i t})\right\}\right].$$

Finally, if $\mathfrak{d}_i t$ is small for all i and $k_{ji}\mathfrak{d}_i s(i)$ is small for each i and j, then (3.14) may be approximated using the equation

$$(3.17) \qquad E[\Delta t(j)] \simeq t(j)\sum_{i=1}^{M} k_{ji}\mathfrak{d}_i s(i).$$

While this expression is definitely of linear law form (compare with (1.15)) the conditions for its validity will not usually obtain in models of theater-level combat. Of the three approximations, (3.16) is definitely the best in theater-level contexts, with (3.15) next best and (3.17) worst.

Despite their being simple (simplistic!) the models presented in this Section are the basis of the attrition calculations in four of the principal models of theater-level combat, as we shall indicate in the next Section.

4. ATTRITION COMPUTATIONS IN THEATER-LEVEL COMBAT MODELS

In this Section we discuss the attrition calculations in four principal computerized simulations of ground-air combat at the theater-level, namely the CONAF Evaluation Model (CEM IV), the IDAGAM I Model, the Lulejian-I Model, and the VECTOR-2 Model. We shall show that the simplified stochastic attritions models described in Section 3 above constitute the core of the calculations in the first 3 of these models (i.e., all except the VECTOR-2 Model) and that the fourth principal theater-level combat models in current use.

We emphasize that the descriptions given in this Section are highly abstracted and that in none of them do we describe in detail the attrition calculations, the explicit nature of effectiveness parameters, or additional dependences on variables such as terrain or posture. Our goal is to indicate the form of the equations used in attrition calculations and to demonstrate how that form is related to the material developed above. For more detailed descriptions of the attrition calculations the reader is referred to the model documentation published by the developers (Louer, 1974; Anderson, et al., 1974; Anderson, Bennett and Hutzler, 1974; Anderson, Blankenship and Karr, 1976; Lulejian and Associates, 1974; Miller, 1977; Miller, et al., 1978), to the author's critiques of the models (Karr, 1976c, 1977b; 1976b; 1977d), on which this Section is largely based, and to the "Four Model Comparison Study" reported in Walker, et al. (1976). A recent report treating the same models is the GAO report (Computer General of the United States, 1980).

All four models are iterative, fixed time step computerized simulations of ground-air combat at the theater level. Each breaks down the combat into (somewhat) smaller-sized interactions in which attrition effects are computed. Furthermore, attrition is generally calculated by applying a unilateral attrition equation first to one side and then (with searchers and targets interchanged and parameter values possibly changed) to the other side. In both calculations the initial numbers of combatants are used as inputs. To illustrate, if for example the homogeneous square law equation (3.2) were used to represent combat between r Red combatants and b Blue combatants, then resulting attritions would be

$$(4.1a) \qquad E[\Delta b] = b\left[1 - \left(1 - \frac{k_1 d_1}{b}\right)^r\right]$$

and

$$(4.1b) \qquad E[\Delta r] = r\left[1 - \left(1 - \frac{k_2 d_2}{r}\right)^b\right],$$

respectively, where d_1, d_2 are detection probabilities and k_1, k_2 are conditional probabilities of kill given attack.

It suffices, therefore, to describe for each model a unilateral attrition equation in terms of targets and searchers (i.e., in the language of Section 3). In passing, we note that none of the models uses asymmetric attrition structures, even though there are interactions, such as that in air-to-air combat between interceptors and escorts, with evident physical and combat objective asymmetries. One way of dealing with this sort of interaction is described in Anderson, Blankenship, and Karr (1976).

Following the notation of Section 3 let

M = number of types of searchers,

$s(i)$ = number of searchers of type i,

N = number of types of targets,

$t(j)$ = initial number of targets of type j.

Let

$$t = \sum_{j=1}^{N} t(j)$$

be the total initial number of targets and, finally, let

$\delta t(j)$ = calculated attrition to targets of type j.

We have introduced the new notation $\delta t(j)$ and terminology "calculated attrition" to emphasize that most of the attrition equations discussed below are inexact, either because they are approximations to begin with or because of difficulties arising from iterative use of them. While we will not discuss this latter problem further, it may be significant and is certainly worthy of some research effort.

Finally, we point out that our notation matches that of Section 3, not those of the model documentations.

We now proceed to describe the attrition calculations in the four models.

4.1. The CONAF Evaluation Model

For the CONAF Evaluation Model (CEM), the situation is simple to describe. Essentially all attrition calculations in this model are (cf. Louer, et al., 1974; Karr, 1976c) performed using equations of the form

$$(4.2) \qquad \delta t(j) = t(j)\left[1 - \exp\left(- \frac{\alpha_j}{t}\sum_{i=1}^{M} p_{ji} s(i)\right)\right] ,$$

where p_{ji} is the "potential of one searcher of type i against targets of type j " and is computed within the model from more primitive and physically oriented parameters, and where α_j is a damage coefficient. The equation (4.2) is the exponential approximation to the heterogeneous square law process of Section 3 that appears as equation (3.12). Observe that there is no direct correspondence between the parameters in (3.12) and those in (4.2). Moreover, the assumptions leading to (3.12) permit one parameter (representing direction or acquisition) depending on searcher type alone and one parameter (representing destruction given an attack) depending on the types of both the search and the target, whereas (4.2) contains one parameter depending on both types and one depending only on the type of the target. This raises the possibility of a conceptual error by the developers of the CEM, which might have occurred because there was no articulation or understanding of the physical assumptions underlying the equation to which (4.2) is an approximation.

4.2. The IDAGAM I Model

We will first discuss the equations used to describe air-to-air combat, and then those used to represent ground combat. Important references are Anderson, et al. (1974), Anderson, Bennett and Hutzler (1974), Anderson (1979) and Karr (1977b).

Air Combat. In the IDAGAM I model there are available a number of attrition equations for calculations concerning air-to-air combat. Although the IDAGAM documentation is not as clear as one would hope in explaining the differences and relationships among the various alternatives, it does at least contain some guidance. The first available choice, and the equation used in most implementations of the model, is

$$(4.3) \qquad \delta t(j) = t(j)\left[1 - \prod_{i=1}^{M}\left(1 - \frac{k_{ji}}{t}[1 - (1-\mathfrak{d})^{t}]\right)^{s(i)}\right] ,$$

where $\mathfrak{d}$ is a detection probability depending on neither the type of the searcher nor the type of the target. Equation (4.3) is an exact equation and is the form to which the heterogeneous linear law equation (3.14) reduces when the detection probability does not depend on the type of the searcher.

Partly for the sake of comparisons but also, in the opinion of the author, somewhat unnecessarily and possibly even misleadingly, the IDAGAM I model contains two approximations to (4.3), namely the equations

$$(4.4) \qquad \delta t(j) = t(j)\left[1 - \exp\left(- \frac{1}{t}\sum_{i=1}^{M} k_{ji} s(i)(1 - e^{-\mathfrak{d}t})\right)\right]$$

and

$$(4.5) \qquad \delta t(j) = t(j)\tilde{d}\sum_{i=1}^{M} k_{ji}s(i) \; .$$

These equations were discussed in Section 3, where they are numbered as (3.16) and (3.17), respectively. As pointed out in Section 3, the two approximations are not equally good; while (4.4) is valid (provided the time interval is short) for parameter values and numbers of searchers and targets likely to be present in theater-level contexts, (4.5) is not.

There is a second attrition fundamentally different from (4.3) that is also available in IDAGAM I as an option for use in calculating air-to-air attrition, and that equation is

$$(4.6) \qquad \delta t(j) = t(j)\left[1 - \exp\left(-\frac{1}{t}\sum_{i=1}^{M} k_{ji}d_i s(i)\right)\right] ,$$

which is the exponential approximation (3.12) to the heterogeneous square law model (3.11) and is nearly the equation used in the CEM. However, the parameter dependences in (4.6), in which the d_i are detection probabilities and the k_{ji} probabilities of target detection given attack, correctly match those in (3.12). Unfortunately, the exact square law equation (3.11) is not available in IDAGAM I, even though the linearized approximation (which is less widely applicable than (4.6))

$$(4.7) \qquad \delta t(j) = \frac{t(j)}{t}\sum_{i=1}^{M} k_{ji}d_i s(i)$$

is available. Of course, (4.7) is precisely the same as (3.13).

If, on the basis of underlying assumptions, one believes that the heterogeneous square law model embodied in (3.11) is an appropriate description of air-to-air combat, then in using IDAGAM I one cannot choose the exact equation, but is forced instead to choose between the two approximations (4.6) and (4.7), which are not (as was discussed in Section 3) equally accurate. At least, however, the available equations are approximations to an exact equation resulting from a known set of assumptions.

Finally, there is also available the equation

$$(4.8) \qquad \delta t(j) = t(j)\left[1 - \prod_{i=1}^{M} (1 - d_i k_{ji})^{s(i)}\right] ,$$

which is an exact heterogeneous version of the Linear Law Process, Multiple Shot Version, discussed in Section 3, with $m = t$, so that every searcher can attack each target it detects. The equation most analogous to (4.8) that appears explicitly in Section 3 is (3.10).

Ground Combat. Ground combat attrition can be calculated in IDAGAM I in two essentially different ways, the first of which will be mentioned only briefly. In the first method potential weapons system losses and actual casualties are calculated separately (the latter from force ratios and historical data); actual casualties are then used to scale potential weapons system losses in order to obtain actual weapons system losses. In this method, the equation used to calculate potential weapons system losses is of the same form as the equation discussed in the next paragraph.

The second method for calculating ground combat attrition uses equations of the form

$$(4.9) \qquad \delta t(j) = \sum_{i=1}^{M} \alpha(j,i) k_{ji} s(i) \ ,$$

where the k_{ji} are probabilities of kill and the $\alpha(j,i)$ are fire allocation factors of the following form:

$$(4.10) \qquad \alpha(j,i) = \frac{t(j)\beta^*(j,i)/t^*(j)}{\sum_{k=1}^{N} t(k)\beta^*(k,i)/t^*(k)} \ .$$

In this expression, $\alpha(j,i)$ is the fraction of searchers of type i that actually fire upon targets of type j and $\beta^*(j,i)$ is the corresponding fraction when the force of targets is a prescribed "standard" force $(t^*(1), \ldots, t^*(N))$. The reader is referred to Karr (1974b) for a proof that all fire allocations satisfying a small set of simple, intuitive axioms are of the form (4.10). By way of illustration, if

$$\beta^*(j,i) = \frac{t^*(j)}{\sum_{k=1}^{N} t^*(k)} \ ,$$

then (4.10) yields

$$\alpha(j,i) = \frac{t(j)}{\sum_{k=1}^{N} t(k)} \ .$$

In other words, if the "standard" allocation is proportional, so is the "actual" allocation.

We note in passing that allocations of the form (4.10) could replace the proportional allocations in equations (2.14), (2.15) and (3.11) above.

Returning to the main exposition, the equation (4.9) is of the form of the approximation (3.13) to the exact heterogeneous square law equation (3.11). The discussion in Section 3 suggests that a corresponding exponential approximation might be preferable.

4.3. The Lulejian-I Model

Attrition calculations in the Lulejian-I model (Lulejian and Associates, 1974) are virtually all performed using equations of the form

$$(4.11) \qquad \delta t(j) = t(j)\left[1 - \exp\left(-\frac{1}{\rho t}\sum_{i=1}^{M}\bar{k}_i s(i)\right)\right],$$

where ρ is a separation distance and where

$$(4.12) \qquad \bar{k}_i = \frac{1}{t}\sum_{\ell=1}^{N} k^*_{\ell i} t(\ell) .$$

The quantity $k^*_{\ell i}$ is the "potential of one searcher of type i to destroy targets of type ℓ," so one should interpret $\bar{k}_i$ as the average destruction potential of one searcher of type i, weighted by the relative numbers of targets present.

The equation (4.11) is the exponential approximation (3.12) to the exact heterogeneous square law equation (3.11), but includes the unnecessary, and in fact incorrect, parameter averaging given by (4.12).

In the three models considered so far, the dominant attrition equation is some form, usually an approximation, of the heterogeneous square law model (3.11); only IDAGAM I provides in some cases the linear law alternative (3.14). Understanding the assumptions leading to (3.11) and (3.14) provides one with essentially complete knowledge of the attrition calculations in the CEM, IDAGAM I and Lulejian-I models.

4.4. The VECTOR-2 Model

Here the equations for ground combat are treated first.

Ground Combat. Attrition calculations for ground combat in the VECTOR-2 model (Miller, 1977; Miller, *et al*., 1978) are in several ways more sophisticated and complicated than those in the other three models, although not without some unique difficulties as well. Not only is the time scale much shorter, which is certain to reduce errors resulting from the very nature of the iterative calculations in all four models, but also there is explicit representation of physical effects such as parallel or serial target acquisition, alternating periods of target visibility and invisibility, random times to acquire targets given continuous visibility, random times to destroy a target given continuous acquisition, and possible destruction of a target by another searcher. In addition to positive consequences, this complexity makes it uncertain whether there exists a definitive set of underlying assumptions. Indeed, since the equations are used to represent manifestly transient phenomena, but are based partly on asymptotic properties of certain stochastic processes, there is some reason to doubt that a set of underlying assumptions exists.

In this paper we cannot discuss in detail either the advantages or the disadvantages of the method of attrition calculation used in VECTOR-2, but we will at least descrive the flavor of it. For details the reader is referred to Karr (1977b), Miller (1977), and Miller *et al* (1978).

Ground combat attrition equations in the VECTOR-2 model are of the general form

$$(4.13)\qquad \delta t(j) = \sum_{i=1}^{M} a(j,i,\underset{\sim}{t},\underset{\sim}{s})s(i) \;,$$

where a is an <u>attrition rate function</u>,

$$\underset{\sim}{t} = (t(1), \ldots, t(N))$$

is the entire target force (type-by-type), and

$$\underset{\sim}{s} = (s(1), \ldots, s(M))$$

is the entire force of searchers. It is tempting, but not justified, to conclude that (4.13) is essentially the deterministic square law model given by (1.10). To do so would be erroneous first of all because the attrition calculations of VECTOR-2 are asserted in Miller (1977) to be fundamentally based on stochastic considerations. Although one can dispute this claim, it does seem reasonable to view (4.13) as a modification of the heterogeneous square law process with generator (2.13). The more fundamental reason why it is incorrect to view (4.13) as merely a square law equation is that the attrition rate functions $a(j,i,\underset{\sim}{t},\underset{\sim}{s})$ depend nontrivially on the target force $\underset{\sim}{t}$ and the searcher force $\underset{\sim}{s}$. Only if there were dependence on neither would (4.13) be precisely a square law equation.

To give an idea of the nature of the attrition rate functions used in the VECTOR-2 model, let us consider the case of searchers of type i that use parallel target acquisition. This means that target types are ranked in order of priority and that when a searcher engages a target it continues to search for targets of higher priority. Should it acquire one it immediately switches its fire to the newly acquired target. The target priorities must be prescribed by the model user and represent nothing physical; the idea that a weapons system might cease engaging a nearby and threatening target because a higher priority target appears in the distance is distinctly unappealing, but that is how the parallel target acquisition scheme operates in VECTOR-2. One then has

$$(4.14)\qquad a(j,i,\underset{\sim}{t},\underset{\sim}{s}) = \alpha_{ji}\left[1 - \left(1 - \frac{\eta_{ji}}{\mu_{ji}+\eta_{ji}}\,\frac{\lambda_{ji}}{\mu_{ji}+\lambda_{ji}}\right)^{t(j)}\right] \times\left[\Pi^*\left(1 - \frac{\eta_{ki}}{\mu_{ki}+\eta_{ki}}\,\frac{\lambda_{ki}}{\mu_{ki}+\lambda_{ki}}\right)^{t(k)}\right] ,$$

where the product Π^* is over all k for which targets of type k are higher priority than those of type j, and where

μ_{ji}^{-1} = mean length of visible periods for one target of type j and one searcher of type i,

η_{ji}^{-1} = mean length of invisible periods for one target of type j and one searcher of type i,

λ_{ji}^{-1} = mean time required for one searcher of type i to acquire one target of type j, given continuous visibility,

α_{ji}^{-1} = mean time required for one searcher of type i to destroy one target of type j, given continuous acquisition.

The three factors on the right-hand side of (4.14) may be interpreted as the kill rate, the probability that a type j target is visible and acquired (and is hence under attack), and the probability that no higher priority target is visible and acquired. We refer the reader to Bonder and Farrell (1970), Karr (1977d) and Miller (1977) for details of the interpretation. Note, however, the following anomly: as $\alpha_{ji} \to \infty$, $a(j,i,\underset{\sim}{t},\underset{\sim}{s}) \to \infty$, which is not sensible. As $\alpha_{ji} \to \infty$, target destruction consumes a negligible amount of time, but target destruction should not occur at an infinite rate since it would still be limited by both invisibility and loss of acquisition.

For descriptions of attrition rate functions for searchers with serial target acquisition we refer the reader to Karr (1977d), Miller (1977) and Miller, et al. (1978).

Air Combat. Attrition in air-to-air combat is calculated in the VECTOR-2 model using equations of the form

$$(4.15) \qquad \delta t(j) = \frac{t(j)}{t} \sum_{i=1}^{M} c(j,i)\min\{t(j),\ m_i\}s(i) \ ,$$

where the $c(j,i)$ are positive constants, interpreted in Miller (1977) as "attrition caused by one [searcher] of type i that engages only [targets] of type j ," and where m_i is the maximum number of targets that a searcher of type i can engage simultaneously. The equation (4.15) is a linearized approximation to one heterogeneous version (there are several heterogeneous versions that make sense) of the "Homogeneous Linear Law Process, Multiple Shot Version" described in Section 3. To fully justify this assertion would require an unreasonable amount of space and additional equations, which simply do not seem really merited at this point in the paper. The unconvinced reader can find some details in Miller (1977) and Miller, et al. (1978). Validity of the approximation (4.15) is questionable in the theater-level context, even for the relatively smaller pieces into which engagements are decomposed in VECTOR-2, and lack of an exact equation and a set of underlying assumptions are significant shortcomings of the model.

To conclude this Section, we observe once again that the simplified attrition models treated in Section 3 include nearly all the equations used in four principal theater-level simulations of ground-air combat. The assumptions set forth in Section 3 contribute much to understanding differences and similarities among the attrition calculations in these four models.

5. CONCLUSION

In this paper we have surveyed several aspects of the theory and applications of Lanchester models of combat, beginning with the original deterministic systems of differential equations, continuing through continuous time Markov process analogues, proceeding to simplified stochastic models, and concluding with a discussion demonstrating that virtually all attrition calculations in four principal ground-air combat models are performed using one of the simplified stochastic models or an approximation to one of the simplified stochastic models. The approach and philosophical basis of this paper are distinctive in seeking to illuminate the nature of the various attrition models in terms of sets of underlying assumptions. Analysis and understanding of assumptions can only lead to more enlightened use of models.

REFERENCES

[1] Anderson, L. B., Bracken, J., Healy, J. G., Hutzler, M. J., and Kerlin, E. P., IDA Ground-Air Model I (IDAGAM I), Volume 1: Comprehensive Description. R-199, Institute for Defense Analyses (Arlington, Virginia, 1974).

[2] Anderson, L. B., Bennett, D., and Hutzler, M. J., Modifications to IDAGAM I. P-1034, Institute for Defense Analyses (Arlington, Virginia, 1974).

[3] Anderson, L. B., Blankenship, J. W., and Karr, A. F., An Attrition Model for Penetration Processes. P-1148, Institute for Defense Analysis (Arlington, Virginia, 1976).

[4] Anderson, L. B., Antipotential Potential. N-845, Institute for Defense Analyses (Arlington, Virginia, 1979). (a)

[5] Anderson, L. B., Attrition Papers Referenced in IDAGAM I. N-846, Institute for Defense Analyses (Arlington, Virginia, 1979). (b)

[6] Bonder, S. and Farrell, R. (eds.), Development of Models for Defense Systems Planning. SRL-2147 TR70-2, Systems Research Laboratory, University of Michigan (Ann Arbor, Michigan, 1970).

[7] Bracken, J., Two Interpretations of Lanchester Processes. Presented at the Conference on Mathematical Models of Conflict and Combat: Theory and Uses (Mt. Kisco, New York, 1979).

[8] Cinlar, E., Introduction to Stochastic Processes (Prentice-Hall, Englewood Cliffs, New Jersey, 1975).

[9] Clark, G. M., The Combat Analysis Model, Ph.D dissertation, Ohio State University, 1969.

[10] Comptroller General of the United States, Report to Congress. Models, Data and War: a critique of the foundation for defense analysis. PAD-80-21, General Accounting Office (Washington, D.C., 1980).

[11] Deitchman, S. J., A Lanchester model of guerrilla warfare, Operations Research 10 (1962) 818-827.

[12] Dolansky, L., Present state of the Lanchester theory of combat, Operations Research 12 (1964) 344-358.

[13] Goldie, C. M., Lanchester square-law battles: transient and terminal distributions, Journal of Applied Probability 14 (1977) 604-610.

[14] Helmbold, R. L., A "universal" attrition model, Operations Research 14 (1966) 624-635.

[15] Jakobovits, R. H., A class of simultaneous point fire attrition models, unpublished memorandum, Institute for Defense Analyses, Arlington, Virginia, 1977.

[16] Jennings, N., The Mathematics of Battle III: approximate moments of the distribution of states of a simple heterogeneous battle. M-7315, Defense Operational Analysis Establishment, U.K. (1973) (a)

[17] Jennings, N., The Mathematics of Battle IV: stochastic linear law battles. M-7316, Defense Operational Analysis Establishment, U.K. (1973). (b)

[18] Karlin, S. and Taylor, H. J., A First Course in Stochastic Processes, 2nd edition (Academic Press, New York, 1976).

[19] Karr, A. F., Stochastic Attrition Processes of Lanchester Type. P-1030, Institute for Defense Analyses (Arlington, Virginia, 1974). (a)

[20] Karr, A. F., A Class of Binomial Attrition Processes. P-1031, Institute for Defense Analyses (Arlington, Virginia, 1974). (b)

[21] Karr, A. F., On Simulations of the Stochastic, Homogeneous Lanchester Square Law Process. P-1112, Institute for Defense Analyses (Arlington, Virginia, 1975). (a)

[22] Karr, A. F., On Simulations of the Stochastic, Homogeneous Lanchester Linear Law Process. P-1113, Institute for Defense Analyses (Arlington, Virginia, 1975). (b)

[23] Karr, A. F., Deterministic Approximations of Some Stochastic Models of Competing Populations. Technical Report 239, Department of Mathematical Sciences, The John Hopkins University, Baltimore, Maryland, 1976. (a)

[24] Karr, A. F., On the Lulejian-I Combat Model. P-1182, Institute for Defense Analyses (Arlington, Virginia, 1976). See also Karr, A. F., Additional Remarks on Attrition and FEBA Movement Calculations in the Lulejian-I Combat Model. P-1277, Institute for Defense Analyses (Arlington, Virginia, 1977). (b)

[25] Karr, A. F., On the CONAF Evaluation Model. P-1210, Institute for Defense Analyses (Arlington, Virginia, 1976). (c)

[26] Karr, A. F., A Class of Lanchester Attrition Processes. P-1230, Institute for Defense Analyses (Arlington, Virginia, 1976). (d)

[27] Karr, A. F., A Review of Seven DOAE Papers on Mathematical Models of Attrition, P-1249, Institute for Defense Analyses (Arlington, Virginia, 1977). (a)

[28] Karr, A. F., On the IDAGAM I Combat Model. P-1265, Institute for Defense Analyses (Arlington, Virginia, 1977). (b)

[29] Karr, A. F., A General Multiple Shot Binomial Attrition Process. WP-1, Project 2371, Institute for Defense Analyses (Arlington, Virginia, 1977). (c)

[30] Karr, A. F., Review and Critique of the VECTOR-2 Combat Model. P-1315, Institute for Defense Analyses (Arlington, Virginia, 1977). (d)

[31] Lanchester, F. W., Aircraft in Warfare: the dawn of the fourth arm. No. V, The Principle of Concentration. Reprinted in J.

Newman (ed.), The World of Mathematics, Volume 4 (Simon and Schuster, New York, 1956).

[32] Louer, P. E., et al., Conceptual Design for the Army in the Field Alternatives Force Evaluation. CONAF Evaluation Model IV. OAD-CR-GO, General Research Corporation (McLean, Virginia, 1974).

[33] Lulejian and Associates, The Lulejian-I Theater Level Combat Model, Volume II: Model Logic and Equations. Report 259, Weapons Systems Evaluation Group (Arlington, Virginia, 1974).

[34] Miller, G. (ed.), The VECTOR-2 Theater-Level Combat Model, Vol. I: Users Manual and Volume II: Program Maintanance Model. VRI WSEG-4, 5, FR-77-1(R), Vector Research, Incorporated (Ann Arbor, Michigan, 1977).

[35] Miller, G., et al., The VECTOR-2 System for Simulation of Theater-Level Combat, I and II. VRI CCTC-1 FR77-1, Vector Research, Incorporated (Ann Arbor, Michigan, 1978).

[36] Morse, P. M. and Kimball, G. E., Methods of Operations Research (MIT Press, Cambridge, Massachusetts, 1951).

[37] Perla, P. P. and Kehoczky, J. P., A New Approach to the Analysis of Stochastic Lanchester Processes: I. Time evolution. Technical Report 135, Department of Statistics, Carnegie-Mellon University (Pittsburgh, Pennsylvania, 1977).

[38] Perla, P. P., Approximation Techniques and Optimal Decision Making for Stochastic Lanchester Models. Technical Report 137, Department of Statistics, Carnegie-Mellon University (Pittsburgh, Pennsylvania, 1978).

[39] Snow, R. N., Contributions to Lanchester Attrition Theory. RA-15078, RAND Corporation (Santa Monica, California, 1948).

[40] Springall, A., Contributions to Lanchester Combat Theory, Ph.D dissertation, Virginia Polytechnic Institute and State University (Blacksburg, Virginia, 1968).

[41] Sternberg, S. R., Development of Optimal Allocation Strategies in Heterogeneous Lanchester Type Processes, Ph.D dissertation, University of Michigan (Ann Arbor, Michigan, 1970).

[42] Taylor, J. G., Lanchester-type models of warfare and optimal control, Naval Research Logistics Quarterly 21 (1974) 79-106.

[43] Taylor, J. G. and Brown, G. G., Canonical methods in the solution of variable coefficient Lanchester-type equations of modern war-fare, Operations Research 24 (1976) 44-69.

[44] Taylor, J. G., Recent developments in the Lanchester theory of combat, in: Haley, K. B. (ed.), Proceedings of the Eighth IFORS International Conference on Operations Research (North-Holland, Amsterdam, 1979).

[45] Taylor, J. G., Force-on-Force Attrition Modeling (Operations Research Society of America, Arlington, Virginia, 1980).

[46] Walker, L. G., Higgins, M. S. and Svetlich, W. G., Comparison and Evaluation of Four Theater-Level Models: CEM IV, IDAGAM I,

Lulejian-I, VECTOR-1. R-299, Weapons Systems Evaluation Group (Arlington, Virginia, 1976).

[47] Watson, R. K., An application of Martingale methods to models of conflict, Operations Research 24 (1976) 380-382.

[48] Weale, T. G., The Mathematics of Battle I: A bivariate probability distribution. M-7129, Defence Operational Analysis Establishment, U.K, 1971. (a)

[49] Weale, T. G., The Mathematics of Battle II: The moments of the distribution of battle states. M-7130, Defence Operational Analysis Establishment, U.K., 1971. (b)

[50] Weale, T. G., The Mathematics of Battle V: Homogeneous battles with general attrition functions. M-7511, Defence Operational Analysis Establishment, U.K., 1975.

[51] Weale, T. G., The Mathematics of Battle VI: The distribution of the duration of battle. M-76126, Defence Operational Analysis Establishment, U.K., 1976.

[52] Weale, T. G. and Peryer, E., The Mathematics of Battle VII: Moments of the distribution of states for a battle with general attrition functions. M-77105, Defence Operational Analysis Establishment, U.K., 1977.

[53] Weiss, H. K., Lanchester-type models of warfare, in: Davies, M, Eddison, R. T., and Page, T. (eds.), Proceedings of the First International Conference on Operational Research (Operations Research Society of America, Baltimore, Maryland, 1955).

MATHEMATICS OF CONFLICT, Martin Shubik (ed.)

DIFFERENTIAL GAMES: CONCEPTS AND APPLICATIONS

Y. C. Ho and G. J. Olsder

Harvard University, Division of Applied Sciences
Cambridge, Massachusetts
Twente University, Department of Applied Mathematics
Enschede, Netherlands

I. INTRODUCTION

The subject of Differential Games has been in existence for over twenty years. A very large body of literature has been built up. An encyclopedic review and survey of the subject is neither possible nor desirable. What we hope to do in this article is to provide a self-contained introduction to the subject, to put into perspective the various ramifications and the development of the subject, and to give a guide to further reading. To accomplish this in reasonable length, we have decided to treat primarily the zero sum and deterministic version of the problem. Even under this restriction, we cannot hope to review all publications. Our effort will concentrate on pointing out the main features and concepts involved with representative references and a clear application bias. Absence of mention of any particular work does not necessarily imply its lack of significance. In particular, we have not been able to assimilate completely the vast amount of differential game literature that have appeared in USSR journals in the past decade. We have attempted, though, to point out the major streams in the Soviet literature and to reference the representative works.[1] To the extent nonzero sum and stochastic differential games represent conceptual extension of the deterministic case, we will discuss their interrelationships. However, no survey on these subjects will be attempted.

We have organized the references into (i) books (indicated by a B after the date of publication), (ii) survey papers and bibliographical sources (indicated by an S), (iii) references cited in this paper and (iv) other references, some of which have been sent ot us by our colleagues, but which are not directly referenced.

II. ZERO-SUM DIFFERENTIAL GAMES

II.1. Fundamentals of Differential Games

The subject of differential games can be approaches in several ways. Historically in the late '50s and early '60s, the subject arose from the study of pursuit and evasion between two objects moving according to simple kinematic laws. Rufus Isaacs, the acknowledged father of DG, singlehandedly created a theory as well as numerous intricate examples illustrating various facets of the pursuit evasion problems. More or less concurrently, control theory which is the theory of one person dynamic optimization problems was given great impetus for advancement from the demands of aerospace guidance and control problems. It was only a matter of time that the conceptual unity between the two subjects became apparent. DG began to be regarded as an extension of control theory. Efforts were devoted to prove rigorously various analogs of major control theoretic results for the two sided case. A certain amount of success was achieved. However, it was also realized that for conceptual generality and applicability DG is best studied as a special subject in game theory. In technical terms, DG, as it is commonly used in the literature, can be characterized as a two person, deterministic, zero sum, almost perfect information, positional game in extensive form. What we have here are the following ingredients:

(i) two players (often called P and E, respectively)[2] and no chance player (hence no randomness in the problem);

(ii) what one player gains in payoff is the loss of the other (zero sum);

(iii) a set of rules which spell out the order of decisions of the two players, the information each player has at the time of each decision, and the evolution of the game as a result of these decisions.

In a general game in extensive form, these rules are completely unstructured and are characterized by an arbitrary and finite game tree. In DG, we require the players to act alternatively in succession (multi-stage). At each decision stage, each player knows all the past decisions of both players except the decision of the other player at the current stage (almost perfect information). Finally, we stipulate that all the decisions of the players in previous stages can be summarized by a state (positional) variable at the beginning of each stage. As a result of these specializations, we can in fact explicitly specify the evolution of the game tree by a set of difference equations

$$x_{t+1} = f(x_t, u_t, v_t); \; x_0 = \text{Given}, \quad t = 0, 1, \ldots, T \tag{2.1}$$

where x_t is the state variable at the start of stage t, u_t and v_t are the decisions of the players during stage t, T is the terminal time prespecified or determined by some appropriate condition. The payoff can now be stated as

$$J = \phi(x_T) \tag{2.2}$$

with u as minimizer and v maximizer and where ϕ is assumed to be monotonic in $|x_T|$. Now consider the class of feedback strategies

$$\Gamma_u = \{k_t(x_t)\,|\,k \text{ any measurable mapping such that } u_t = k_t(x_t)\}$$

and similarly for Γ_v and $v_t = g_t(x_t)$. We define the DG problem as

$$\text{(DG)} \qquad \min_{k\in\Gamma_u} \max_{g\in\Gamma_v} J \quad \text{or} \quad \max_{g\in\Gamma_v} \min_{k\in\Gamma_u} J$$

Associated with Γ_u we can define another class of strategies

$$\Gamma'_u = \{k_t(x_t, v_t)\,|\,k : X_t \times V_t \to U_t\} \tag{2.4}$$

In other words, in this case u_t has the advantage that at each stage he acts after v_t knowing the choice of v_t. Similarly we define Γ'_v. Then we have two associated games

(DG-upper) $\quad \underset{k \in \Gamma_u}{\text{Min}} \; \underset{g \in \Gamma'_v}{\text{Max}} \; J$

(DG-lower) $\quad \underset{g \in \Gamma_v}{\text{Max}} \; \underset{k \in \Gamma'_u}{\text{Min}} \; J$

Since both (DG-upper) and (DG-lower) are finite perfect information games of extensive form, the existence of value and saddle point strategy pair for these games are guaranteed (Owen, 1968). Furthermore, since $\Gamma_u \subset \Gamma'_u$ and $\Gamma_v \subset \Gamma'_v$, it can be shown (Danskin, 1976)

$$J(\text{DG-lower}) \le J(\text{DG}) \le J(\text{DG-upper}) \tag{2.5}$$

where $J(\cdot)$ stands for the saddle point, if it exists, of the game. This is intuitively plausible since we argue that in a zero sum situation more information never hurts a player; he can always disregard the new information and guarantee his payoff under the old information structure by definition. Now if we replace (2.1) by its continuous time analog

$$\frac{dx}{dt} = f(x,u,v,t) \; ; \quad x(t_0) = \text{Given} \tag{2.6}$$

and generalize (2.2) to

$$J = \phi(x(t_f)) + \int_{t_0}^{t_f} L(x,u,v,t)dt \tag{2.7}$$

then (DG) represents what we commonly accept as "Differential Game" when the admissible strategy class Γ_u and Γ_v are also appropriately modified. Thus, continuous time DG can be regarded as the limit of multistage game when the discretization interval approaches zero. It is also reasonable to argue that under certain conditions the difference $J(\text{DG-upper}) - J(\text{DG-lower})$ may approach zero as we shrink the discretization interval since the informational advantage of one game over other may become vanishingly small. The natural condition for equality to be attained in (2.5) turns out to be the

<u>Separability Assumption</u>:

$$f(x,u,v,t) \equiv f_1(x,u,t) + f_2(x,v,t)$$

$$L(x,u,v,t) \equiv L_1(x,u,t) + L_2(x,v,t)$$

Because of this requirement, the minimum principle in control theory can be replaced by a minimax principle since the separability assumption guarantees that $\underset{u}{\text{Min}}\,\underset{v}{\text{Max}}\, H = \underset{v}{\text{Max}}\,\underset{u}{\text{Min}}\, H$ where the Hamiltonian $H \equiv L + \lambda^T f$. In this connection, the solution of (DG) under (2.6) and (2.7) can be expected to obey the Hamilton-Jacobi-Bellman-Isaacs Equation

$$-\frac{\partial J}{\partial t} = \underset{u}{\text{Min}}\ \underset{v}{\text{Max}}\left[L + \frac{\partial J}{\partial x} f\right] \ ; \quad J(x(t_f)) = \phi(x(t_f)) \tag{2.8}$$

The solution of (2-8), $J^0(x,t)$, is called the value of the DG. This is again in accordance with well known results of control theory.

II.2. Theoretical Development

A considerable part of DG theory is directed towards the rigorous establishment of the validity of (2.8), the attainment of equality in (2.5) under various assumptions on the limiting process, and the definition of the value of the DG. The important references here are Fleming (1961, 1964), Varaiya and Lin (1969), Roxin (1969), Friedman (1971B), Elliott (1972, 1977a). The basic problems here are two fold. First, a careful delineation of the strategy spaces of the two players P1 and P2. This is necessary so that we can guarantee the existence of the solution of the differential equation $\dot{x} = f(x,u,v,t)$ when u and v take on these strategies. Given the resultant $x(t)$ we can then evaluate the payoff J which depends on $x(t)$, $u(t)$ and $v(t)$. To accomplish this, we often discretize the time and endow one or the other strategy space with an information advantage in terms of what a player knows at the time of decision. (Re: Eq. (2.4).) These definitions of strategy result in two possible values for J. The second problem then addresses the convergence of these two values under some suitable limiting process of the time-steps approaching zero.

To model the information flow both Fleming (1961) and Friedman (1971B) approximate the differential game (2.6), (2.7) by a discretized version; the time interval $[t_0, t_f]$ is partitioned into k subintervals of equal length. At the beginning of an interval $[t_i, t_{i+1}]$, P1 (the minimizer) chooses a control u_i on $[t_i, t_{i+1}]$, with the knowledge of the previous u_j and v_j, $j < i$. Subsequently, P2 chooses v_i, based on the same knowledge and in addition u_i. Thus P1 is thought of to play first at each time t_i. The best the two players can obtain with this information pattern is the upper-value W_k^+. The lower-value W_k^- is defined similarly; now P2 has to play first on each subinterval. For $k \to \infty$ the values will be indicated by W^- and W^+, provided these limits exist. In the limit of the first game described, we can write $u(t) = \gamma_1(x(t), t)$, $v(t) = \gamma_2(x(t), t, u(t))$, and in the second game $u(t) = \gamma_1(x(t), t, v(t))$, $v(t) = \gamma_2(x(t), t)$. The difference between Fleming and Friedman is that in Fleming constant controls were considered on the subintervals, whereas in Friedman the controls on the subintervals are measurable.

The approach in Varaiya, Lin and Roxin starts with the classes U_1 and U_2 of admissible, measurable control function u and v on the interval $[t_0, t_f]$. The lower value U^- is defined as

$$U^- = \sup_{\alpha} \inf_{u_1 \in U_1} J(u, \alpha(u)) \ ,$$

where α, which is nonanticipative, maps U_1 into U_2. The upper-value U^+ is similarly defined. The approach has further been developed in Elliott (1972). It has been shown that $W^- = U^- = U^+ = W^+$, provided that the so-called Isaacs condition holds, which is that

$$\min_u \max_v (p'f + L) = \max_v \min_u (p'f + L)$$

for all n-vectors p. This condition equals the separability condition mentioned earlier.

Other works related to rigorous definitions of strategies are Krasovskii and Subbotin (1977B), Danskin (1973), and Pontryagin (1967a). In Krasovskii a discussion of the differential equation $x = f(x, \gamma_1(x,t), \gamma_2(x,t), t)$ is given when the strategies $\gamma_i(x,t)$ are not necessarily Lipschitz in x. A solution of such a differential equation is defined, which is now referred to as a Krasovskii-solution. For other solution concepts of ordinary differential equations the reader is referred to Hajek (1979).

Blaquiere and Leitmann, independent of Isaacs, derived the Hamilton-Jacobi-Bellman-Isaacs equation (2.8) (Blaquiere, Gerard and Leitmann, 1969B; Blaquiere and Gerard, 1968; Leitmann and Wang, 1969; Blaquiere, 1971). Their (geometric) approach is essentially the same as Isaacs', but is more rigorously stated. Extensions to differential games with time-lags are reported in Blaquiere (1972), Ciletti (1970a), Blaquiere and Caussin (1973); the role of mixed and behavioral strategies in such games is specifically considered by Petrosyan (1970) and Levine (1979). Extensions to differential games with piecewise continuous trajectories, including impulsive controls, are treated in Blaquiere (1977). See also Korneev and Melikian (1977). Also some work has been done with respect to differential games for which the underlying system is described by a PDE instead of an ODE (Roxin, 1977; Bensoussan, 1971). A connection of the HJBI equation with wave mechanics is indicated in Blaquiere (1980). In Blaquiere (1976b) time and state dependent coalitions are considered in many player differential games.

II.3. Solution Approaches

For the actual solution of DG, main stream game theory is of very little help since few results are known for games in extensive form. We must turn to results from control theory. To do this we must examine the statement of the DG problem more carefully. Given a dynamic systems (2.6) and a cost function (2.7) satisfying the separability assumption we wish to find a strategy pair $u = \gamma_1^*(x,t)$, $v = \gamma_2^*(x,t)$ such that

$$J(\gamma_1^*, \gamma_2^*) \leq J(\gamma_1, \gamma_2^*) \quad \forall \gamma_1 \in \Gamma_u \tag{2.9a}$$

$$J(\gamma_1^*, \gamma_2^*) \geq J(\gamma_1^*, \gamma_2) \quad \forall \gamma_2 \in \Gamma_v \tag{2.9b}$$

The first question is what should be the admissible class of strategies, Γ_u and Γ_v. It is tempting and reasonable to this point to let Γ be the class of all measurable functions of x and t on account of our (positional) assumption. Note that for any fixed $v = \gamma_2^*(x,t)$, Eqs. (2.6), (2.7), and (2.9a) define a control problem from the viewpoint of player P1 so long as the differential equation $\dot{x} = f(x, u, \gamma_2^*(x,t), t)$ has properly defined solutions over the interval of control. Furthermore, suppose we are successful in "explicitly" solving this control problem defined by (2.9a) and obtain for the optimal control law for player P1 $u = \gamma_1^0(x, t; \gamma_2^*)$ parametrized by γ_2^*. Now we can fix $u = \gamma_1^0$ and solve the similar control problem Eq. (2.9b) faced by player P2. Assuming we are also successful in this and obtain $v = \gamma_2^0(x, t; \gamma_1^0)$, the consistency requires

$$\gamma_2^*(x,t) \equiv \gamma_2^0(x, t; \gamma_1^0(x, t; \gamma_2^*)) , \tag{2.10a}$$

$$\gamma_1^*(x,t) \equiv \gamma_1^0(x, t; \gamma_2^0(x, t; \gamma_1^*)) . \tag{2.10b}$$

Resolution of the implicit equations (2.10) then furnishes us with saddle point strategy pairs. However, this is not the whole story. First, despite the fact the game is positional, there is no a priori reason not to consider strategies of this type, e.g., $v = \gamma_2(x(\tau), t_0 \le \tau \le t)$. If such strategies were used, then the control problem faced by player P2 will in fact be a much more difficult one involving functional differential equations. In fact, we have generally no assurance that another saddle point does not exist in the class of strategies involving memory for this position. What rescues us from this dilemma is the fact that this problem is zero sum. Suppose another saddle point γ_1^0, γ_2^0 exists which involves memory, then we must have $J(\gamma_1^*, \gamma_2^*) = J(\gamma_1^*, \gamma_2^0) = J(\gamma_1^0, \gamma_2^*)$ by virtue of the results from control theory in connection with the two control problems solved above. But by definition of (γ_1^0, γ_2^0), $J(\gamma_1^0, \gamma_2^*) \le J(\gamma_1^0, \gamma_2^0) \le J(\gamma_1^*, \gamma_2^0)$ which imply that all four values must be equivalent and all strategies interchange (Owen, 1968B). Consequently, once (γ_1^*, γ_2^*) is determined, the question of uniqueness becomes unimportant. It is for this reason that we can safely restrict the search to Γ. See also Basar (1977a). Note, however, that this argument is NOT applicable when the admissible Γ is not rectangular (e.g. certain strategy pairs may not be playable against each other). Thus, if termination of the game requires the satisfaction of some conditions such as the state belong to certain target manifolds in the state space, then not all strategy pairs may be playable (Leitmann, 1974). In fact a large body of DG literature is devoted to the study of conditions to guarantee the satisfaction of these terminal conditions. Since these terminal conditions often require the capture of one player by another, they come under the name of capturability. The associated DG is known as game of kind (vs. degree).

A second difficulty appears in the iterated solution process described above. We are asked to solve explicitly for $\gamma_1^0(x, t; \gamma_1^*)$ and $\gamma_2^0(x, t; \gamma_1^*)$ i.e. obtain the feedback control law. This is rarely possible in general except for special classes of problems such as those with linear dynamics and quadratic payoff (Bryson and Ho, 1969B) or for problems of low dimensions. To circumvent this difficulty we appeal to the principle of backward induction (or optimality) or dynamic programming. Instead of fixing the control laws for v <u>for all</u> t (a long term view) and then iterate as we have described above, we can fix the control law for only the last stage or instant of control. We then go through the iteration process exactly as described for u and v only for this last stage. This is a short term problem. Particularly, since the problem is completely deterministic, γ_1^* and γ_2^* and hence the payoff, can in fact often be explicitly determined for one stage of a game problem in this fashion when dimension of the state x is small (say 2 or 3). The principle of backward induction then enables us to repeat this process for other stages and instances one at a time. Almost all explicit solutions of DG problems are accomplished in this manner (Isaacs, 1965B). See also Section III for more details. What makes this process of solution intellectually interesting and nontrivial is the large number of "singular" surfaces that are continuously being discovered in example problems through this solution process and the insight they offer. These singular surfaces are manifolds in the state space across which Eq. (2.8) cannot be continued, i.e., the backward solution process of dynamic programming fails. These singular surfaces occur when the operation $\underset{u}{\text{Min}}\ \underset{v}{\text{Max}}\left[L + \frac{\partial J}{\partial x} f\right]$ results in nonunique saddle points and/or $\partial J/\partial x$ which is discontinuous. Such occurrences are quite frequent since it is well known that the operation of "max" or "min" on smooth functions often results in non-smooth functions. There are generally no counterparts of these surfaces in control theory.

More generally, one may attempt to solve DG problems as two-sided optimal control problems using various standard numerical algorithms for dynamic optimization. This approach is sound so long as only open loop equilibrium (saddle point) solutions are required. Under appropriate conditions, such open loop solutions can yield useful information to help the synthesis of (sub)optimal feedback strategies. While such an approach is difficult to justify on theoretical grounds, one may adopt it as a last resort in practice. See for example Anderson and Bohn (1977), Anderson (1977), Taylor (1975) and Jarmark (1976).

II.4. Other Conceptual Extensions

Once we have defined differential games as in (DG) and Eqs. (2.6), (2.7), certain natural generalizations suggest themselves. Two main directions for extensions exist. First, we can relax the assumption of zero sum. Thus two players may have payoff functions

$$J_1 = \phi_1(x(t_f)) + \int_{t_0}^{t_f} L_1(x, u_1, u_2)dt \quad \text{and} \quad J_2 = \phi_2(x(t_f)) + \int_{t_0}^{t_f} L_2 dt$$

with $J_1 \neq J_2$. Furthermore, there need not be only two players but N players with payoffs J_i, $i = 1, \ldots, N$. However, once we step outside the bounds of the zero sum assumption the full complexity of main stream game theory presents itself. Issues such as core, coalitions, pareto-optimality and threats must be investigated. To some extent, such effort has been successful and provided research avenues to pursue. But by and large, majority of the work has been on the <u>noncooperative</u> version of n-person differential games. In this context, the natural analog of (DG) is

$$J_i(\gamma_1^*, \ldots, \gamma_n^*) \leq J(\gamma_1^*, \ldots, \gamma_{i-1}^*, \gamma_i, \gamma_{i+1}^*, \ldots, \gamma_n^*) , \quad i = 1, \ldots, n \tag{2.11}$$

where $u_i = \gamma_i^*(x,t)$ and $\dot{x} = f(x, u_1, \ldots, u_n, t)$. Eq. (2.11) is known as the Nash Equilibrium Solution of a Nonzero Sum Differential Game. Many real world problems, particularly economic problems, can be formulated as NZSDG where the interests of the players are not in direct conflict. However, to date relatively few results have been obtained and there are little or almost no analogs of the plethora of "singular" surfaces in zero sum cases, an exception is Sarma and Prasad (1976).

A variation of the pursuit-evasion problem concerns the case of two pursuers vs. one evader. From nonzero sum game viewpoint this corresponds to the special case where $J_1 = J_2 = -J_3$. More generally, we can visualize two teams of players engaged in a zero sum problem. With such a setup, many of the N-person game theoretic complications can be by-passed, one can then concentrate on the intricacies of the problems that are unique with differential games.

The second direction of generalization for DG introduces the presence of uncertainty into the problem. Following standard control theoretic setup we modify (2.6) to

$$\dot{x} = f(x, u_1, u_2, w, t) \tag{2.12}$$

$$z_i = h_i(x, v_i, t) \tag{2.13}$$

where $w(t)$ and $v_i(t)$ are some well defined stochastic processes. Let $\zeta_i(t)$ represent the history $\{z_i(\tau), t_0 \leq \tau \leq t\}$. We now define

$$\Gamma_i = \{\gamma_i(\zeta_i(t), t) | \gamma_i : \zeta_i \times T \rightarrow U_i\} \tag{2.14}$$

i.e. the strategies are functions of the players' past observation history. Then a stochastic DG can be stated as

$$\underset{\gamma_1 \in \Gamma_1}{\text{Min}} \; \underset{\gamma_2 \in \Gamma_2}{\text{Max}} \; J \tag{2.15}$$

Various conceptual difficulties arose early with the problem of Eq. (2.15) all of which are traceable to the lack of precision and understanding in the definition of Γ_i and the role of information structure in extensive games. For example, since different players have different observation histories there was early concern with the problem of "second guessing" (namely, "I'll estimate your estimate of my estimate of your estimate...."). Also if one player builds a state estimator based on assumed optimal strategy of the other player then what happens to the estimate if the other player no longer plays optimally. Computationally, difficulties also crop up due to the nature of the admissible strategy set Γ . The problem of dealing with functional differential equation and infinite dimensional strategy discussed earlier can no longer be avoided. Basically, we have a situation in which one's information depends on the strategy of the other, namely what is known as dynamic information structure. The conceptual and computational issues involved are very intricate and we refer interested readers to a separate paper (Ho and Chu, 1974). Suffice it to say, very little is known about solutions of stochastic DG beyond the Linear-Quadratic-Gaussian case. Even in that special case, resolution of the various above-mentioned conceptual issues took some time.

III. PURSUIT-EVASION WITH PERFECT STATE MEASUREMENT

In this section we will be concerned with the DG defined by (2.6) and (2.7). The information available to the players will be the current state and time, hence the name perfect state measurement. "Pursuit-evasion" in the section's title has been used since most games to be treated here will have a clear pursuit-evasion interpretation.

In III.1 we deal with linear-quadratic games, which have attractable analytical solutions. Research in this direction started with Ho, Bryson and Baron (1965). Before addressing the problem of how well the players can act, e.g. how soon the pursuer can catch the evader, one should first consider the more basic problem of whether capture is at all possible. This is the subject of subsection III.2. Subsection III.3 deals with linear games, i.e. the state dynamics is linear. The Pontryagin-approach also has been briefly mentioned in this subsection. Lastly, III.4 deals with singular surfaces. Some examples are given, such as the homicidal chauffeur.

III.1. Linear-Quadratic Games

Consider two dynamic systems being situated in the Euclidean space R^n

$$\begin{aligned} \dot{x}_p &= A_1 x_p + B_1 u_1 , \\ \dot{x}_e &= A_2 x_e - B_2 u_2 , \end{aligned} \tag{3.1}$$

where the subscripts p and e stand for pursuer and evader respectively. The vectors u_1 and u_2 are r_1, respectively r_2-dimensional. The pursuer uses the control u_1 to attempt to capture the evader, whereas the evader tries to avoid capture. The measure of capture is defined as the miss-distance at a given final time t_f :

$$\| x_p(t_f) - x_e(t_f) \|^2_{Q_f} \tag{3.2}$$

where the weighting matrix Q_f is symmetric and > 0 . For both players there will be a trade-off between the miss-distance and the actual amount of energy used, here defined as

$$\int \| u_1 \|^2_{R_1} dt \ , \quad \int \| u_2 \|^2_{R_2} dt$$

respectively. We again assume $R_i' = R_i$, $R_i > 0$, $i = 1, 2$. The pay-off function is therefore defined as

$$J = \|x_p(t_f) - x_e(t_f)\|^2_{Q_f} + \int_0^{t_f} \{\|u_1\|^2_{R_1} - \|u_2\|^2_{R_2}\}\, dt \tag{3.3}$$

By redefining the above stated problem in terms of $x \triangleq x_p - x_e$, we see that this problem is a special case of the following, so-called linear-quadratic, differential game:

$$\min_{u_1} \max_{u_2} J, \quad J = \tfrac{1}{2}x'(t_f)Q_f x(t_f) + \tfrac{1}{2}\int_0^{t_f} \{x'Qx + u'R_1 u - u_2'R_2 u_2\}dt \tag{3.4}$$

subject to

$$\dot{x} = Ax + B_1u_1 + B_2u_2 , \quad x(0) = x_0 , \quad 0 \leq t \leq t_f . \tag{3.5}$$

We will look for solutions of the form $u_1 = \gamma_1(x,t)$, $u_2 = \gamma_2(x,t)$.

Several methods exist to solve this DG. Assuming the value function to be of the form

$$V(x,t) = \tfrac{1}{2}x'S(t)x \tag{3.6}$$

turns out to be a winning bet. Substitution of (3.6) into (2.8) leads to the requirement that the $n \times n$ matrix S must satisfy

$$\dot{S} = -Q - A'S - SA + S(B_1R_1^{-1}B_1' - B_2R_2^{-1}B_2')S , \quad S(t_f) = Q_f \tag{3.7}$$

If the solution to this matrix differential equation exists, then the optimal strategies are

$$u_1(t) = -R_1^{-1}B_1'S(t)x(t) , \quad u_2(t) = +R_2^{-1}B_2'S(t)x(t) \tag{3.8}$$

Another quick way to obtain this solution is to "complete the square" in the integrand of (3.4). It is readily verified that

$$J - \tfrac{1}{2}x_0'S(0)x_0 = \tfrac{1}{2}\int_0^{t_f} \Big\{(u_1 + R_1^{-1}B_1'S(t)x)'R_1(u_1 + R_1^{-1}B_1'S(t)x)$$
$$+ (u_2 - R_2^{-1}B_2'S(t)x)'R_2(u_2 - R_2^{-1}B_2'S(t)\Big\}dt ,$$

where $S(t)$ satisfies (3.7), which also leads to the saddlepoint strategies (3.8).

The popularity of linear-quadratic DGs is on the one hand because they arise in the perturbation analysis of nonlinear pursuit evasion game problems (with final end time) and on the other they are analytically tractable. The value function is twice differentiable in the state vector (no singularities!) and the optimal strategies are easy to implement since they are of the linear feedback type.

A sufficiency condition for the existence of a saddlepoint of this game is the existence of the solution of Eq. (3.7) which is guaranteed if either R_1^{-1} and R_2^{-1} are sufficiently small for a fixed t_f the quadratic term in S in (3.7) does not contribute significantly) or $B_1R_1^{-1}B_1' \geq B_2R_2^{-1}B_2'$ for any t_f. In the latter case there exists a uniform bound M such that $\|S(t)\| < M$ for all t. If neither of these conditions are fulfilled, backward integration of (3.7) may lead to an explosion of $S(t)$, i.e. $\lim_{t\downarrow\tau}\|S(t)\| = \infty$ for a certain τ. In terms of the calculus of variations one has arrived at a conjugate point (or more properly a focal point). It has been shown in Bernhard (1979), that absence of a conjugate point is not a necessary condition for the existence of saddlepoint strategies.

Though the linear quadratic DG has been stated with constant matrices A_1, B_1, B_2, R_1, R_2, Q, the results remain valid if they are piecewise continuous functions of time. For linear quadratic DGs with $t_f = \infty$ the reader is referred to Pachter (1978), Mageirou (1976), and Jacobson (1977). It has been shown there that one must be careful when applying the finite horizon theory and taking the limit $t_f \to \infty$.

Though there do not seem to exist too many applications, the theory is often used as a starting point for linear-quadratic games with imperfect state observations or noisy measurements, see Section V.2.

III.2. Termination and Capturability

In this section we will consider games for which the final time is not fixed *a priori*, but is implicitly given by some termination-conditions. The actual final time will in general depend on the actions of the players. A closed set Λ, called the target set, is assumed to be given in the production space of the time-axis and the state space. The final time is defined as

$$t_f = \min\{t \mid (x(t), t) \in \Lambda\} . \qquad (3.9)$$

Sometimes the terminal time is defined slightly differently, viz. as the smallest time for which $x(t)$ penetrates Λ (Bernhard, 1977). For our purposes (3.9) will suffice. The boundary of Λ, $\partial\Lambda$, is assumed to be given by a scalar function $\ell(x,t) = 0$, thus forming an (n-1)-dimensional manifold in R^n. The target set Λ consists of those points for which $\ell(x,t) \leq 0$.

Before addressing the problem of finding a saddlepoint of the cost function (2.7), one must consider the question whether the target will at all be reached. If not, the game problem (2.6), (2.7) is not well defined. In this section we will be concerned with a more basic game: the evader tries to prevent the state from reaching Λ whereas the pursuer's goal is opposite. One could define an auxiliary cost function as: $J = -1$ if the state reaches Λ and $J = +1$ if it does not. Once one knows that Λ can be reached, one can address again the original problem as defined by (2.6), (2.7).

Only those points (x,t) of $\partial\Lambda$ for which

$$\min_{u_1}\max_{u_2} \nu' f(x, u_1, u_2, t) = \max_{u_2}\min_{u_1} \nu' f(x, u_1, u_2, t) \leq 0 \qquad (3.10)$$

are candidates for a possible terminal situation of a game. The n-dimensional vector ν is the projection of the outward normal at Λ at (x,t) on X. If strict inequality holds then the state will penetrate Λ. Points (x,t) for which the equality sign holds can possibly only be grasing-points. All points satisfying (3.10) are together called the usable part (UP) of $\partial\Lambda$.

Consider Figure 1. The shaded region consists of all initial points from where the differential game will terminate. How do we find the

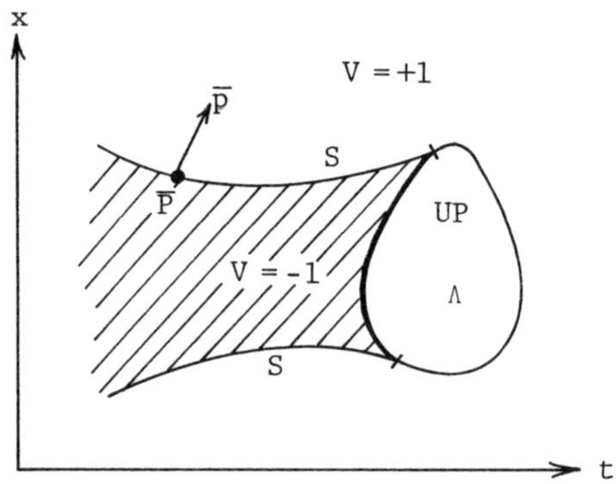

FIGURE 1

surface P which separates those initial points from the others from where termination is not possible? Consider an arbitrary point $\overline{P}$ of P. The outward normal is indicated by $\overline{p} = (p, p_{n+1})$, where p is n-dimensional. A necessary and sufficient condition that P separates the points from where termination is/is not possible, is that

$$\min_{u_1} \max_{u_2} p'f(x, u_1, u_2, t) + p_{n+1} = 0 . \tag{3.11}$$

Because of this separation property, P is called a semipermeable surface; without the pursuer's help the evader cannot make the state cross P (from the shaded to unshaded region) and vice verse, without the evader's help, the pursuer cannot make the state vector cross P (from the unshaded to the shaded region).

By substituting the minimizing u_1 and maximizing u_2 as functions of x and t into (3.11) an identity in x and t is obtained. Differentiation w.r.t. x of this identity leads to Isaacs (1965B)

$$\frac{dp}{dt} = -\left(\frac{\partial f}{\partial n}\right)' p , \quad \frac{dp_{n+1}}{dt} = -\left(\frac{\partial f}{\partial f}\right)' p . \tag{3.12}$$

From (2.6), (3.11) and (3.12) the surface P can be in principle determined. As final conditions for the differential equations (2.6) and (3.12) we take those x (and corresponding normal ν) of $\partial\Lambda$ for which the equality sign holds in (3.10).

For time-independent problems, i.e. problems for which f , ℓ (and the cost function) do not explicitly depend on t , the analysis can be pursued a little further. The analog of Figure 1 then is given by Figure 2a, where, for simplicity, a system with only two state components, x_1 and x_2 , will be considered.

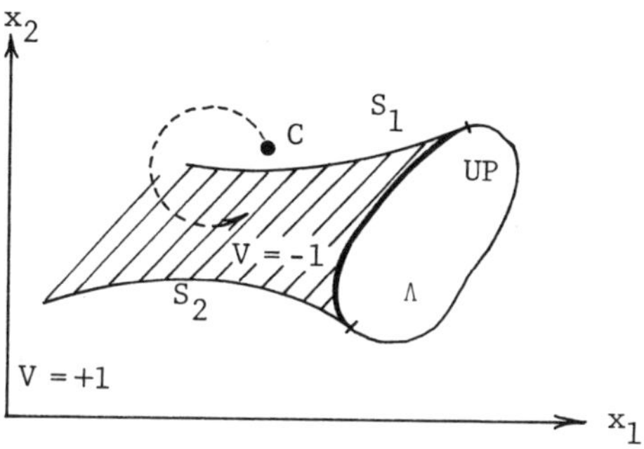

FIGURE 2a

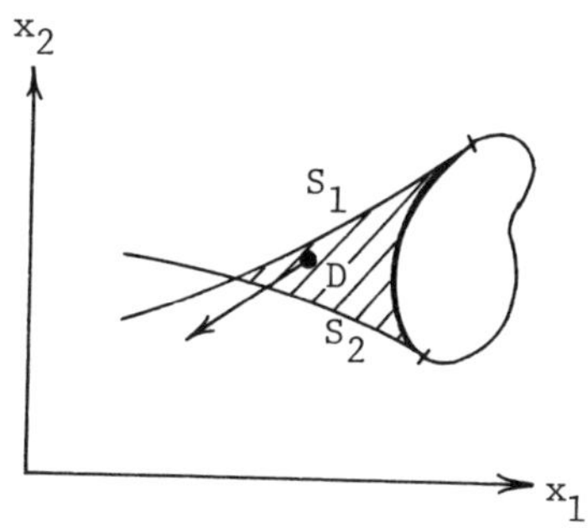

FIGURE 2b

The whole state space can be divided into two parts: initial points from where P can force termination and initial points from where E can enforce escape Λ forever. A rigorous proof of this statement is given in Krasovskii and Subbotin (1977B). Extensions in various directions are given, see for instance Maksimov (1976a, 1976b), Chikrii (1975), and Kleimenov (1975). We start from this intuitively appealing hypothesis. If the shaded area in Figure 2a is separated from the nonshaded area, P_1 and P_2 have to intersect. P_1 and P_2 are sometimes denoted by BUP; boundary of the usable part. Thus P_1 , P_2 and UP will enclose the points from which termination can be ensured. If P_1 and P_2 do not intersect, but are of finite length, then termination from every point in state space is possible. If the initial point is at C , see Figure 2a, then P cannot force the state vector cross P_1 without E's help, he can however, cause the state vector to go around P_1 , as indicated by the dashed line.

Also if the lines P_1 and P_2 intersect, it is not yet guaranteed that they separate the areas with $V = +1$ and $V = -1$. In addition to intersection one has to require that the point of intersection does not "leak." One can envisage the situation of P keeping the state at his "right" side, i.e., the shaded side, of P_1 , but thus being forced to let the state vector cross P_2 . This has been illustrated in Figure 2b. In such a situation E can always enforce escape. One speaks of a leaking corner. Though in Figure 2b, the leaking is to E's advantage, one can easily envisage examples where the leaking is to P's advantage. The problem of closed and leaking corners has been considered in Bernhard (1971).

Let us consider two examples. The first one is the famous "homicidal chauffeur"-game. Let us consider a pursuer and evader who move in a two-dimensional plane according to

$$\dot{x}_{1p} = v_1 \sin \theta_p \,, \quad \dot{x}_{2p} = v_1 \cos \theta_p \,, \quad \dot{\theta}_p = \omega_1 u_1 \,, \quad |u_1| \le 1 \,,$$

$$\dot{x}_{1e} = v_2 \sin \theta_e \,, \quad \dot{x}_{2e} = v_2 \cos \theta_e \,, \quad \dot{\theta}_e = \omega_2 u_2 \,, \quad |u_2| \le 1 \,.$$

where v_1 and v_2 are the constant velocity of P and E respectively. The maximum angular velocities are ω_1, resp. ω_2. If $u_p - 0$ $(u_e = 0)$, P(E) will move in a straight line; $u = +1(-1)$ refers to the sharpest possible turn to the right (left). The curvature of the trajectories is bounded. By defining relative coordinates as

$$x_1 = (x_{1e} - x_{1p})\cos \theta_p - (x_{2e} - x_{2p})\sin \theta_p \,,$$

$$x_2 = (x_{1e} - x_{1p})\sin \theta_p - (x_{2e} - x_{2p})\cos \theta_p \,,$$

$$\theta = \theta_e - \theta_p$$

we get the following model:

$$\dot{x}_1 = -\omega_1 u_1 x_1 + v_2 \sin \theta$$

$$\dot{x}_2 = -v_1 + \omega_1 u_1 x_1 + v_2 \cos \theta$$

$$\dot{\theta} = \omega_2 u_2 - \omega_1 u_1 \qquad (3.13)$$

We introduced this model, sometimes referred to as the two-car model for later reference. In the homicidal chauffeur game, the second player can change his direction instantaneously, i.e. $\omega_2 = \infty$. Put differently, E has complete control over θ, which will therefore be taken as his control. Also normalizing P's velocity and its maximum angular velocity to one, we get therefore

$$\dot{x}_1 = -u_1 x_2 + v_2 \sin u_2 \,,$$

$$\dot{x}_2 = -1 + u_1 x_1 + v_2 \cos u_2 \,, \qquad (3.14)$$

with no bounds on u_2 and u_1 satisfies $|u_1| \le 1$. In this model the pursuer's velocity-vector is aligned with the positive x_2-axis. Figure 3 shows the system in relative coordinates. The reason for going into some detail w.r.t. this model is that many authors used it (Isaacs, 1965B; Breakwell, 1973; Merz, 1971), see also subsection IV.1, it is very versatile in the sense that together with suitable cost functions the corresponding game shows many singular surfaces, and moreover, some of them will be indicated in this survey. Termination occurs when

$$x_1^2 + x_2^2 = \beta^2$$

where it is assumed that initially $x_1^2(0) + x_2^2(0) > \beta^2$. The pursuer tries to get within a distance β of the evader, whereas the evader tries to avoid this. The game thus formulated is known as the "homicidal chauffeur."

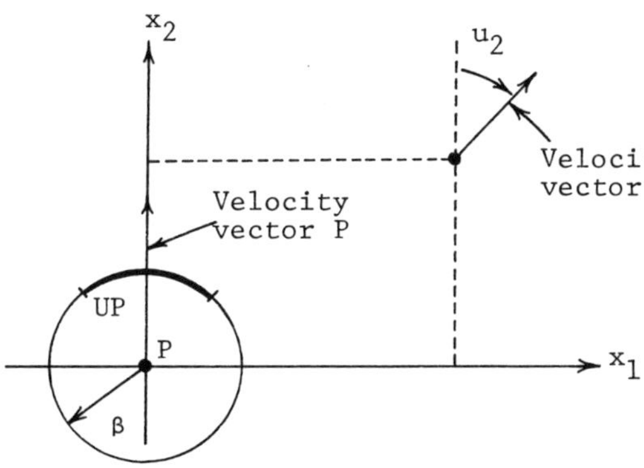

FIGURE 3

The UP, part of the circle $x_1^2 + x_2^2 = \beta$ (being a cyclinder in the (x,t) space), is determined by those x_1, x_2 for which (note that at the point (x_1, x_2), the vector $\begin{pmatrix}\nu_1\\ \nu_2\end{pmatrix} = \begin{pmatrix}x_1\\ x_2\end{pmatrix}$ is an outward normal to Λ):

$$\min_{u_1}\max_{u_2}\{x_1(-u_1x_2 + v_2 \sin u_2) + x_2(-1 + u_1x_1 + v_2 \cos u_2)\}$$

$$= \min_{u_1}\max_{u_2}\{-x_2 + v_2(x_1 \sin u_2 + x_2 \cos u_2)\}$$

$$= -x_2 + v_2\sqrt{x_1^2 + x_2^2} = -x_2 + v_2\beta \leq 0$$

Hence $x_2 \geq v_2\beta$ on the circle $x_1^2 + x_2^2 = \beta^2$ forms the UP. Intuitively this is clear. Directly in front of P, E cannot escape. If E is close to P's "side," he can side-step, and being behind P, i.e. $x_2 < 0$ there is no danger of immediate capture at all.

It can be shown (provided that $\beta^2 + v_2^2 < 1$), see Isaacs (1965B), that P_1 and P_2, being the semipermeable surface starting from $x_2 = v_2\beta$, $x_1 = \pm\beta\sqrt{1 - v_2^2}$, either intersect (on the x_2-axis) and form a closed corner at the point of intersection or they do not intersect and have finite length, depending on the parameters β and v_2. In the latter case capture from any initial state in the space is possible.

The second example is a differential games extension of the brachistochrone problem in the calculus of variation. Isaacs called it the dolichobrachistochrone. The system is described by

$$\dot{x}_1 = \sqrt{x_2}\cos u_1 + \frac{w}{2}(u_2+1)$$

$$\dot{x}_2 = \sqrt{x_2}\sin u_1 + \frac{w}{2}(u_2-1) \ . \tag{3.15}$$

where w is a positive constant. There are no constraints on u_1, u_2 must be chosen such that $|u_2| \leq 1$. The initial condition lies in the first quadrant, $x_i(0) > 0$. The game terminates when $x_1(t_f) = 0$. Only trajectories will be considered for which $x_2(t) \geq 0$ throughout the game. The UP is determined by

$$\min_{u_1}\max_{u_2}\left(\sqrt{x_2}\cos u_1 + \frac{w}{2}(u_2+1)\right) = -\sqrt{x_2} + w \leq 0$$

and is thus given by $x_1 = 0$, $x_2 \geq w^2$. (Note that $\nu_1 = +1$, $\nu_2 = 0$). Some analysis, using (3.11) and (3.12), yields that the equations for the BUP, starting from $x_1 = 0$, $x_2 = w^2$, are:

$$x_1(t) = -\frac{w}{2}(t_f - t) + \frac{w^2}{2}\sin\left(\frac{t_f - t}{w}\right),$$

$$x_2(t) = \frac{w^2}{2}\left(1 + \cos\left(\frac{t_f - t}{w}\right)\right) . \tag{3.16}$$

To our dissatisfaction we notice that $x_1(t) < 0$ for t near t_f and that the BUP as determined by (3.16) is not as indicated as in Figure 2a; here it moves <u>inside</u> Λ ! Isaacs calls this a "discouraging possibility," the trajectory (3.16) clearly cannot fulfill the

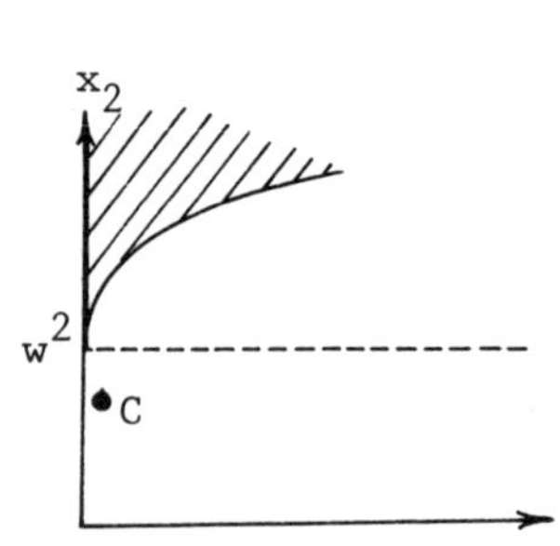

FIGURE 4a

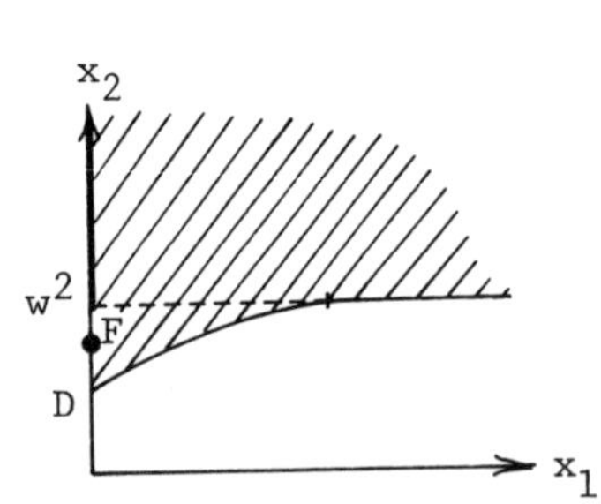

FIGURE 4b

role we wished it to do. Apparently no BUP of the form as indicated in Figure 4a exists. The explanation of what really happens has been given in Breakwell (1971) and Chigir (1976). If the state is in point C, see Figure 4a, then P cannot force it to the half plane $x_1 < 0$, because E can always make $\dot{x}_1 > 0$ by playing $u_2 \approx +1$. However, if E plays $u_2 \approx +1$, P can play $u_1 = \pi/2$ (instead of $u_1 = \pi$ which does not lead to $x_1 < 0$) and the state moves up, i.e. $\dot{x}_2 > 0$. E could keep the state from moving up by playing $u_2 \approx -1$ but then P can force termination. Hence E cannot prevent the state from moving up if he does not want the state to cross the x_2-axis directly. In this way the state moves up until $x_2 > w^2$ and then P can easily finish the game. All initial points from where P can enforce termination form the shaded area in Figure 4b. The explicit formulas are given in Chigir (1976).

Once the problem of termination has been solved, we return to the original DG of the form (2.6), (2.7). For initial points for which P (or E if we put the responsibility of termination with E) cannot enforce termination, this game is not well-defined, unless one adds some extra rules to the game. One possibility is to restrict the class of strategy-pairs (γ_1, γ_2) to such pairs which lead to termination in finite time (Blaquiere, Gerard, Leitmann, 1969B). Another possibility is to include non-termination in the cost function. If the game does not terminate in finite time then $J = \infty$ (puts the responsibility of termination with the minimizing player) or $J = -\infty$ (puts the responsibility of termination with the maximizing player). The latter approach has for instance been adopted in Krasovskii and Subbotin (1977B).

A different approach to capture has been treated in Cockayne (1967). Simple (necessary and sufficient) conditions are given under which capture is possible. The players move in a plane with constant speed and they have upperbounds on the curvatures of their paths. See also Cockayne and Hall (1975).

III.3. Linear Differential Games and the Pontryagin Approach

Another approach to capture has been developed in Pontryagin (1967a, 1967b) and subsequently in Hajek (1975B), for differential games, where the dynamics is given by a linear differential equation. The idea is that the pursuer, who wants to steer the state to the target set, now called capture set, first neutralizes the evader's actions and uses the left-over potential to steer to the capture set. In order to neutralize the opponent's actions, the pursuer must have access to his opponent's current decisions which may or may not be a realistic assumption. Suppose the evolution of a system is described by

$$\dot{x} = Ax - \bar{u}_1 + \bar{u}_2 , \quad x(0) = x_0 \tag{3.17}$$

where x, $\bar{u}_1, \bar{u}_2 \in R^n$, $\bar{u}_1$ and $\bar{u}_2$ are the decision variables of P and E respectively, which we assume to satisfy $\bar{u}_1 \in U_1$, $\bar{u}_2 \in U_2$ for all t involved. With (3.17) we associate the control system

$$\dot{y} = Ay - u \ , \quad y(0) = x_0 \tag{3.18}$$

a one-player game where u must satisfy $u \in U_1 \overset{*}{-} U_2$. The difference $U_1 \overset{*}{-} U_2$ is called the Pontryagin difference of the two sets U_1 and U_2 and is defined as

$$U_1 \overset{*}{-} U_2 = \{u \mid u + U_2 \subset U_1\} \ .$$

Assuming that U_1 and U_2 are convex, symmetric around the origin and nonempty, it can be shown that $U_1 \overset{*}{-} U_2$ is nonempty if and only if $U_2 \subset U_1$ (Hajek, 1975B). If the pursuer knows E's current decisions for all t involved, instead of (3.17) we may well consider (3.18). The latter system is simpler since control theory has been developed much further than the theory of DG's. The pursuer can force the state $x(t)$ to the capture set Λ if and only he can steer $y(t)$ to this set. Thus he faces a controllability-problem. Pontryagin's approach is slightly more general than described here; the generalization being that P knows E's action up to a given number of time units in the future. In that set up, cancellation of E's control by P's control needs not be instantaneous. An extension has been dealt with in Gamkrelidse and Kharatishviti (1974). See also Bernhard (1972), Pontryagin (1974a, 1974b), Petrosyan (1972), Gusiatnikov (1976), Al'brekht and Loginov (1976), Pshenichnyi (1968), and Rechtschaffen (1979).

As an example, consider the system described by

$$\ddot{x} = u_1 + u_2$$

where x, u_1 and u_2 are scalars. This system, of which the description can easily be converted in (3.17), denotes a point with mass one on which two players can apply forces. The forces are restricted to

$$|u_1| \le 3 \ , \quad |u_2| \le 2$$

P's goal is to steer the system to $x = \dot{x} = 0$ whereas E wants to postpone this moment as long as possible. If we assume that P knows E's current decisions, then we can consider the optimal control problem of steering y to $y = \dot{y} = 0$ as soon as possible subject to

$$\ddot{y} = -u \ , \quad |u| \le 1 \ .$$

This latter problem is a standard example in optimal control theory and can be found in any textbook on this subject.

In Heymann, Pachter and Stern (1976a, 1976b) and Heymann and Pachter (1979) the concept of max min controllability has been introduced (as a generalization of controllability in system theory) and its properties investigated. Roughly speaking, max-min controllability makes precise the concept of one player being more powerful than another in terms of controllability (energy required to drive the system to a desired state).

III.4. Singular Surfaces and Examples

In the section on linear-quadratic games, the value function was twice (continuously) differentiable in the state-vector. In this section we will consider problems where the corresponding value function does not have such a smooth behavior. In particular, value functions can be discontinuous (if the discontinuity occurs along an (n-1)-dimensional manifold in state space we speak of a barrier) and/or $\partial V/\partial x$ can be discontinuous.

An assumption almost always made at the outset of any problem is that the state space, or at least the "interesting" part of the state space, can be split up into a finite number of mutually disjoint regions, the value-function being twice differentiable in each of these regions. The boundaries of these regions are called "singular surfaces." The behavior and methods of construction of $V(x,t)$ from eq. (2.8) are well-understood in such regions. A difficulty that remains is to "join" the value-functions from the different regions. Another problem is how to find the singular surfaces; they do not necessarily automatically show up by integrating (2.8) backward in time starting from the target set. Before we present an example showing this, we will describe a method of obtaining $V(x,t)$ from (1.7) in a region where it is twice differentiable. For simplicity's sake we assume the game-problem (2.6), (2.7), together with the target set to be time-independent, which also makes V time-independent, and thus (2.8) reduces to

$$\operatorname*{Min}_{u_1}\ \max_{u_2} \left[L + \frac{\partial V}{\partial x} f\right] \equiv 0 \ . \tag{3.19}$$

Substitution of minimizing u_1 and maximizing u_2, to be denoted by u_1^* and u_2^* respectively, we obtain

$$H(x^*,\ u_1^*,\ u_2^*) \triangleq L(x^*,\ u_1^*,\ u_2^*) + \frac{\partial V(x^*)}{\partial x} f(x^*,\ u_1^*,\ u_2^*) \equiv 0 \tag{3.20}$$

where x^* is the corresponding trajectory. Differentiation of H w.r.t. x yields

$$\frac{\partial L}{\partial x} + \frac{\partial V}{\partial x}\cdot\frac{\partial f}{\partial x} + \frac{\partial}{\partial x}\left(\frac{\partial V}{\partial x}\right) f \equiv \frac{\partial L}{\partial x} + \frac{\partial V}{\partial x}\cdot\frac{\partial f}{\partial x} + \frac{d}{dt}\left(\frac{\partial V}{\partial x}\right) \equiv 0 \ ,$$

where the arguments x^*, u^*, v^* have been omitted and where we already substituted $\partial H/\partial u_1 \cdot du_1/dx = \partial H/\partial u_2 \cdot du_2/dx = 0$, the latter equalities following from (3.19). Thus, along an optimal path,

$$\frac{d}{dt} V_x = -L_x - V_x f \ , \tag{3.21}$$

where $V_x = \partial V/\partial x$, etc.

As an example consider the system described by

$$\dot{x}_1 = (2+\sqrt{2})u_2 - \sin u_1 \ ,$$

$$\dot{x}_2 = -2 - \cos u_1 \ ,$$

with an initial condition in the half plane $x_2 > 0$. The target set is $x_2 \leq 0$ and the cost function is

$$J = x_1^2(t_f)$$

No restrictions on u_1, u_2 satisfies $|u_2| \leq 1$. It turns out that every point on the x_1-axis can be a terminating point. The H-function is

$$H = V_{x_1}\left\{(2+\sqrt{2})u_2 - \sin u_1\right\} + V_{x_2}\left\{-2 - \cos u_1\right\}, \tag{3.22}$$

where, according to (3.21)

$$\dot{V}_{x_1} = 0 \,, \quad \dot{V}_{x_2} = 0 \,. \tag{3.23}$$

Since V is the value-function, we obviously have $V_{x_1} = 2x_1$ along the x_1-axis. Minimaxing (3.23) yields

$$u_2 = \mathrm{sgn}(V_{x_1}) = \mathrm{sgn}(x_1(t_f)) \,, \tag{3.24}$$

$$(\sin u_1, \cos u_1) \parallel (V_{x_1}, V_{x_2})$$

where $\parallel$ stands for parallel to. Substitution of (3.24) and $V_{x_1}(x(t_f)) = 2x_1(t_f)$ into H yields the identical zero-function, from where $V_{x_2}(x(t_f))$ can be solved resulting in $V_{x_2}(x(t_f)) = 2|x_1(t_f))|$. Thus we obtained the terminal conditions for (3.23). In this way we find for terminal conditions $x_1 > 0$, $x_2 = 0$, the constant strategies

$$u_1^* = \frac{\pi}{4} \,, \quad u_2^* = 1$$

and for terminal conditions $x_1 < 0$,

$$u_1^* = -\frac{\pi}{4} \,, \quad u_2^* = -1 \,.$$

(Please note that u_1 is measured clockwise from the positive x_2-axis.) Drawing the corresponding trajectories we obtain Figure 5, and we note that the trajectories intersect. Starting from an initial point I_1, there are two possibilities. Obviously, the trajectory going direction "south-east" is optimal. Starting from I_2, "south-west" is optimal. The separation-line between going south-east and north-west is obviously the x_2-axis. Thus integrating the optimal paths backwards in time from the end-conditions, we should stop this integration once we have reached the x_2-axis. The x_2-axis is called a dispersal line. More precisely a dispersal line for the evader, since if the initial condition is on the x_2-axis, the u_2-player can determine the direction in which to go. For a more geometrical derivation of this solution, see Isaacs (1965B).

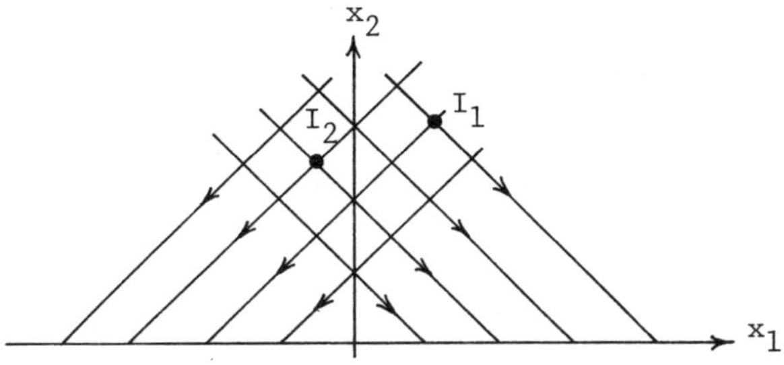

FIGURE 5

A dispersal line is only one of the many singular surfaces. It also occurs in the homicidal chauffeur. Suppose that in this game termination is always possible and that the cost function is the time necessary for capture. If the evader is far behind the pursuer, i.e. in Figure 3 its position is $x_1 = 0$, $x_2 << -\beta$, then P will turn either right or left, as soon as possible in order to get E in front of him. It can rigorously be proven that the half line $x_1 = 0$, $x_2 \leq \alpha$ for some negative α , is a dispersal line for the pursuer. By means of the same game we will introduce another type of singular surface, vis. the so-called universal line. Suppose that in Figure 3 the evader is situated far in front of P , $x_2 << \beta$, relatively close to the x_2-axis. The best P can do is to turn toward E , such that in the relative (x_1, x_2) space, E's position will move toward the x_2-axis. Once E is on the x_2-axis, the remaining part of the game is a trivial game consisting of a chase along a straight line. A line, or more general surface, at which optimal trajectories enter from both sides and then stay on it, is called a universal line (surface).

In Figure 6, we have sketched the dispersal line, the universal line, together with some other singular surfaces. In contrast to these other singular surfaces, the universal line is special in so far that both V and $\partial V/\partial x$ are continuous across the line. The difference between a universal line and focal line is that in the latter case the trajectories enter tangentially, whereas in the previous case they do not. An analogous difference exists between equivocal line and switching envelope. For the latter two surfaces the optimal strategies are not unique. One of the players can decide as to stay on the singular surface for a while or not.

Some of the singular surfaces can only be approximated by strategies of the form $u_1 = \gamma_1(x)$, $u_2 = \gamma_2(x)$. In case of the focal line for instance, one player, say P , is trying to get the state vector off the focal line, but E will bring the state back to the focal line. Strategies of the form $u_1 = \gamma_1(x)$, $u_2 = \gamma_2(x(t), u_1(t))$, with an informational advantage for the evader, would be able to keep the state on the focal line precisely.

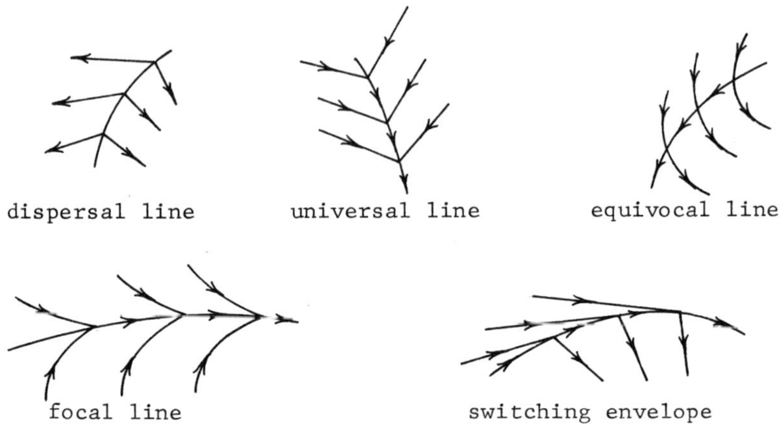

FIGURE 6

We will not treat these surfaces in detail here, but refer to them later on, when dealing with specific examples. The first three singular surfaces on Figure 6 have extensively been considered in Isaacs (1965B), the latter two of Figure 6, appeared for the first time in Breakwell (1973) and Merz (1971).

In all these cases $V(x)$ is continuous across the singular surfaces, but $\partial V/\partial x$ is not, except for the universal surface. Another kind of singular surface exists, called a barrier, across which $V(x)$ itself is discontinuous. As an illustration, let us consider the homicidal chauffeur game again with time of capture as cost function. In the previous section, we have indicated how the semipermeable surfaces starting from the endpoints of the UP of the target could be constructed. If they do not intersect, they will have finite length; they are sketched in Figure 7.

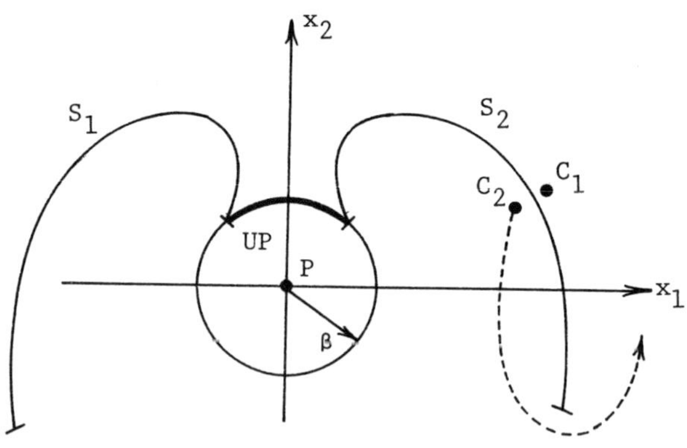

FIGURE 7

These surfaces are barriers. The value function $V(x)$, denoting the minmax capture time when the initial state is x, is discontinuous across P_1 and P_2. If E is initially situated at C_2, P cannot force him across P_1, but he can force him to go around the barrier P_1. Therefore it is plausible--and it can be proven rigorously--that capture from C_2 will take more time than capture from C_1.

A crucial question in the construction of $V(x)$ is how do we know where the singular surfaces are located? Up to now this question has not been answered satisfactorily. It is almost an art to construct them. Once a function $V(x)$ has been constructed, such that it is twice differentiable in each of a finite number of mutually disjoint regions of the state space, there exists sufficiency-conditions to check whether the function $V(x)$ really represents the value function (Bernhard, 1977). Theorems dealing with these sufficiency conditions are called verification theorems after Isaacs. Note that checking the optimality of one strategy against another is only a control-theoretic question for which a great deal is known.

Though termination may be guaranteed, i.e. from every initial point the player responsible for termination can indeed enforce termination, this may be in conflict with his optimal strategy as explained by the following example, which is due to Bernhard (undated).

The kinematics is

$$\dot{x}_1 = u_1 + u_2 \,, \quad -1 \le u_1, u_2 \le 0 \,, \quad x(0) > 0 \,,$$

where x, u_1, and u_2 are scalars. The target set is $x \le 0$. Obviously either player can terminate the game. Suppose the cost function is

$$J = \int_0^{t_f} (1-x)\,dt \quad \text{if} \quad t_f < \infty \,,$$

$$J = \infty \qquad\qquad \text{if} \quad t_f = \infty \,, \tag{3.25}$$

i.e. P is responsible for termination. Using the necessary conditions of this section, we get as candidates for the optimal solution

$$u_1 = \gamma_1^*(x) = \begin{cases} 0 & \text{if} \quad x > 1 \\ -1 & \text{if} \quad x \le 1 \end{cases} \,,$$
$$u_2 = \gamma_2^*(x) = \begin{cases} -1 & \text{if} \quad x > 1 \\ 0 & \text{if} \quad x \le 1 \end{cases} \,, \tag{3.26}$$

which leads to $V(s) = \frac{x(1-x)}{2}$. However, if $x_0 > 1$, E does much better by playing $\gamma_2^*(x) \equiv 0$; if P sticks to $\gamma_1^*(x)$ from (3.26) then the cost function becomes ∞ since $t_f = \infty$: Therefore (3.26) does not contribute a saddlepoint solution. Giving an information advantage to P in the form of $u_1(t) = \gamma_1(x(t), u_2(t))$, where

$$\gamma_1(x(t),\ u_2(t)) = \begin{cases} -1 & \text{if } x \leq 1 \text{ or } (x < 1 \text{ and } u_2 = 0) \\ 0 & \text{otherwise} \end{cases}$$

will restore the equilibrium. This feature of playability versus non-termination also cleared and explained the so-called bang-bang-bang surface introduced by Isaacs (1969; Ciletti, 1970b; Lewin, 1976). A bang-bang-bang surface was introduced as a surface at which each player's optimal strategy is discontinuous, even so this did not follow from the necessary conditions (3.19). Further examples:

In this subsection, we will consider examples, which can be stated rather simply and for which the analysis of solution is in general surprisingly complex. The importance of these examples is to show certain features such as singular surfaces, though they were originally motivated by practical applications. In the next section, we will consider examples in which the application stands more central. This split-up of examples is somewhat arbitrary; the boundary between the two classes is not sharp.

The homicidal chauffeur has been introduced in the previous subsections. The problem originated from Isaacs (1965B) and was partly solved here. The complete solution is quite complicated and is spread out over Merz (1971) and Breakwell (1973), it contains dispersal surfaces, equivocal lines, universal lines, barriers and switching envelopes. See also Pachter and Yavin (1980) for a stochastic version of the homicidal chauffeur. Introduction of uncertainties "smoothens" the singular surfaces which will therefore no longer exist as much.

The dolichobrachistochrone also originated from Isaacs (1965B) and a partial solution has been given there. The complete solution can be found in Breakwell (1971) or Chigir (1976). It contains a singular surface not treated so far, called "safe contact." The problem has been introduced in this paper in Subsection III.2. The cost function will be the time of capture, i.e. the time needed for the system to reach the x_2-axis. The safe contact appears between two points D and F of the x_2-axis, see Figure 4b, and is only properly defined with the information structure $u_1 = \gamma_1(x)$, $u_2 = \gamma_2(x, u_1)$. P wants to move the state in the positive x_2-direction, along the x_2-axis as soon as possible, E tries to prevent this. The informational advantage to E is needed in case P suddenly tries to "pull" the state to the left half plane, in which case the game has terminated. Properly speaking, the "safe contact" should happen just right of the x_2-axis, since termination was defined as $x(t_f) \leq 0$, or we should have changed the definition of termination into $x_1(t_f) < 0$.

The lady in the lake is a very well known example. The solution is for instance given in Breakwell (1977). In this example a young lady is swimming in a circular lake with a maximum velocity v_0 which is constant. The initial position is the center of the lake. She can change the direction in which she swims instantaneously. A man with obscure intentions, who has not mastered swimming, is on the side and can run along the shore with maximum velocity 1. He can change his direction instantaneously, too. The lady does not want to stay in the lake forever and wants to escape; on land she can run faster than the man. It turns out that $v_0 \geq .217...$ the lady can always escape from the man. A special feature of this game is that if the lady makes a mistake, this does not necessarily work to our disadvantage, since she can always return to the center of the lake and start all over again. A related game is the lion and man (Breakwell, 1977; Flynn, 1974) where the lion (pursuer) and man (evader) are both confined to a circular arena. The lion, having a slower maximum speed than the man, tries to come as close as possible to the man, who on his turn tries to maximize the minimum distance between him and the lion. A surprising result is that if the lion starts at the center of the arena and the man is on the perimeter of the arena, the man has first to run towards the lion.

The deadline game considers two players, P and E, in the plane, both with no restrictions on the manoevability and P's maximum speed slightly greater than E's maximum speed. The game is played in the half plane $x_2 \geq 0$. The question is whether E can pass between P and the x_1-axis without coming closer to P than a specified distance β. The answer to this "game of kind" is obviously obtainable if we can solve the following related "game with terminal payoff": What is the distance of closest approach β from any starting position, assuming that P minimizes β and E maximizes β while passing between P and the x_1-axis. See Breakwell (1975) for the solution.

A related game is patrolling a channel. Here E has to pass P, at either side, but now the motions are confined to a strip $0 \leq x_2 \leq x_{2max}$. See Breakwell (1977).

Other games in the same class with melodious names are: Cornered rat (Breakwell, 1977), Obstacle Tag (Breakwell, 1971), Suicidal Pedestrian (Hagedorn and Carmele, 1978). The Wall-pursuit game (Isaacs, 1965B), the Isotropic rocket (Bernhard, 1970), the Two-cars game (Merz, 1972). Other games of the same class can be found in Isaacs (1965B), Bernhard (1975), Breakwell (undated) and Masle (1976), they do however not have standard names attached to them.

Problems with a slightly different origin are the surveillance games. The pursuer wants to keep the evader forever in a given region (say within a certain distance from himself), and if this is not possible, keep E there as long as possible (Liu, 1977; Breakwell and Lewin, 1975; Lewin, Olsder, 1979).

Problems with two pursuers, respectively two evaders have been treated in Hagedorn and Breakwell (1976), Breakwell and Hagedorn (1979), and Kelley (1973). See also for many player-extensions: Foley and Schmitendorf (1976), Chernous'ko (1976) and Chikrii and Rappoport (1979).

IV. APPLICATIONS

Aside from the general theoretical development of DG as briefly discussed in Subsection II.2, the study of singular surfaces can essentially be viewed as qualitative applications. The overall purpose is to make precise the role of various parameters (such as the speed ratio vs. the turning radius in the homicidal chauffeur problem) in the outcome of an abstracted and idealized version of a real world problem. The basic limitation of such an approach to the study of dynamic models of conflict are two fold. First, the model must be of low dimension involving no more than 3 state variables. The study of singular surfaces is crucially dependent on our ability to draw and to visualize geometrical relationships. Secondly, most analysis of singular surfaces are carried out via ad hoc techniques on a case-by-case basis. No easy generalizations are possible. Consequently, DG applications are primarily qualitative. Analysis where applicable serves to define the main features or structures of real world conflict situation. In this section, we shall review further applications of the various qualitative ideas developed earlier to situations where DG as narrowed defined are not directly applicable but where these ideas can serve as a guide or to provide partial insights.

IV.1. Role Determination

A potential field of application is "role determination," sometimes also referred to as two-target games. A dog fight problem, in which each player tries to destroy his opponent before he himself is destroyed, consists of two parts. Firstly, given the kinematics and the initial conditions, determine the roles of the two players; who can destroy whom, i.e. who will be called the pursuer? Secondly, given the roles of the players, what are the best strategies, e.g. time optimal strategies, for the players? In problems of "role determination" only the first question is considered. Destruction is defined as the entry of the state in a given target set (e.g. weapon-range). Obviously, each player wants the state to enter his target set before the opponent's target set is entered. Hence the name "two-target-game." If we assume that a stalemate cannot arise (which would occur, e.g. if the players circle around each other and if a derivation from the strategy for any of the players would lead to his destruction, thus they are forced to keep circling forever), the state space must be divisible into regions corresponding to victory by one or the other player, or to escape by the faster player (if he cannot win). Such a division was qualitatively described in the last sections of Baron, Ho and Kleinman (1970). It should be emphasized that in a deterministic differential game with perfect sight, i.e. perfect state measurements, the roles of the players are completely determined by the initial conditions and will not change during the game, provided of course that the pursuer does not make mistakes.

The region R_1 of the state space corresponding to victory by player P1, is bounded by a surface consisting of (part of) the target set Λ_1 of terminal states defining victory for P1, together with a semi-permeable surface P_1, which prevents the state from leaving region R_1, provided P1 plays optimally in the neighborhood of P_1. The interior of R_1 does not contain any points of the terminal set Λ_2 corresponding to victory by player P2 and thus, assuming that the game ends in final time, victory for P1 is guaranteed for initial states inside R_1. The region R_2 of the state space $(R_1 \cap R_2 = \phi)$ corresponding to victory by player P2 is similarly defined. If R_1 and R_2 do not fill the whole state space, then there are initial points from where neither player can guarantee his victory. This latter region, to be denoted by $R_{neutral}$, could for instance be the region from where the faster player can escape.

This situation is illustrated schematically in Figure 8.

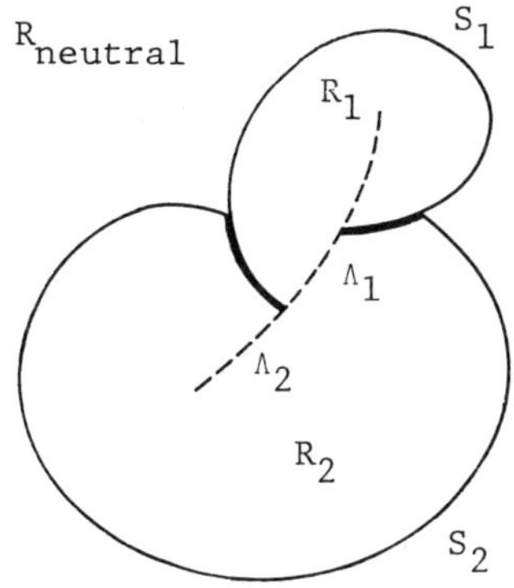

FIGURE 8

How can the semi-permeable surfaces P_1 and P_2 be constructed? Two cases can be distinguished: (i) the part of P_1 that is also part of P_2 (drawn thick in Figure 8), and (ii) the parts of P_1 or P_2 that separate R_1 respectively R_2 from $R_{neutral}$. The first case refers to "simultaneous kill" and the second to "near miss"--situations, i.e., one player "almost" destroyed the other but was not in a dangerous situation himself. For the construction of R_1 and R_2 one first has to recognize the final conditions of the state vector which correspond to either simultaneous kill or near miss situations and secondly one determines the semi-permeable surfaces starting from these final conditions.

In Olsder and Breakwell (1974), this approach has been carried out for two aircraft which move in a single plane and of which the kinematics is described by (3.13). The "mathematical" state space thus is three dimensional. The target set for either player consists of a line segment of length ℓ_i $(i = 1, 2)$ in the direction of its velocity vector. It is shown that there are several types of simultaneous kill and near miss, each of them giving rise to a different semi-permeable surface. Parts of these surfaces taken together form the composite semi-permeable surfaces P_1 and P_2 as indicated in Figure 8, but should here be thought of as a two dimensional manifold in R^3.

In Merz (1975, 1976), the same problem is considered, the only difference being that different target sets are considered. The problem treated in Vincent, Cliff, Grantham and Peng (1973) is slightly different in the sense that the question is not who pursues whom, but rather can one of the players, called the evader, avoid a collision with the other player? If one considers a worst case situation, this latter player will pursue the evader, which exactly is the situation considered in the paper. The state space has been split up into three regions: (i) green; the state of the two aircraft in such that if the evader continues his current direction (with constant speed), collision is not possible; (ii) yellow; the evader can escape a collision only if he executes a collision avoidance maneuver, and (iii) red; despite any maneuver by the evader, collision is possible. See also Peng and Vincent (1975), Vincent, Cliff, Grantham and Peng (1974), Vincent, Sticht and Peng (1976), and Vincent (1975).

It is very hard to extend the techniques of constructing semipermeable surfaces, that enclose a certain area, to a state space of dimension higher than 3. The reason is that it is very difficult to visualize when or to give conditions under which a composite semipermeable surface really encloses an area, i.e. there are no "holes" in the composite surface. Moreover checking whether all corners don't leak won't be an easy task. No attempts in this direction are known, in spite of the fact that model (3.13) does not include vertical movements of the airplanes.

The techniques of constructing semipermeable surfaces between regions corresponding in different outcomes of the game, has also been applied to ship collision avoidance problems. The target set is then not a weapon range, but the shape of the ship itself. Hence "near miss" situations do not exist in such a case; only "simultaneous kill" exists. Large super tankers cannot change their speed rapidly and are not very well maneuverable, reason why a model of the form (3.13) may not be too unrealistic.

In Merz (1973) model (3.13) is used; also some background and motivations are given there. The problem solved is under which conditions one ship can avoid collision with another, identical ship (i.e. same shape, velocity, maximum angular velocity) which (i) also tries to avoid the manevrer, (ii) does not maneuver, but instead travels in a straight path on the surface of the sea. Both cases do not actually belong to the field of differential games, since there is not a conflict. In Vincent and Peng (1973) the ship collision avoidance problem is considered from a worst case design point of view and as in Vincent, Cliff, Grantham and Peng (1973), they construct green, yellow and red regions. See also Vincent (1977). Part of their solution, however, has the "discouraging possibility" of the semipermeable surface at the wrong side of the target, which has not been resolved. This feature has been discussed in Section III.2 in the dolichobrachistochrone problem.

In Sharma (1976), Miloh and Sharma (1976, undated) also ship-collision avoidance is considered with three different roles for each player; evasive, pursuing and neutral (i.e. travelling straight) thus leading to nine different cases. Various modelling aspects are discussed. As in Merz (1973) and Vincent and Peng (1973), a collision takes place if the ships come closer to each other than a given distance. This means that the ships are considered to be round discs, as viewed from above. In Olsder and Walter (1978) the target sets are line-segments, which are more ship-shaped. This leads, however, to a greater analytical complexity, since the "boundary" of a line segment is not smooth at its ends.

In Getz and Pachter (1980a) a role determination is considered with respect to the simplified 2-dimensional model (3.4). They, however, also give strategies (not necessarily time-optimal) which will lead to termination in finite time within each region. The mathematical techniques include Lyapunov functions. See also Getz and Leitmann (1979), Pachter and Getz (1980) and Getz and Pachter (1980b).

A different approach with respect to role determination has been taken in a series of papers that start with Baron, Ho and Kleinman (1970). In this approach the game is modeled discrete time and discrete state and the evaluation of the state can be described by a Markov-chain. The action spaces are necessarily discrete too, such that the evolution of the state matches with the state-discretization. In principle the roles could be determined by playing the game backward in time, as we did in the continuous set-up. In Falco and Cohen (1973) another method has been followed to obtain optimal strategies and a division of the state space into several regions, each corresponding to a particular outcome of the game. This method offers the conceptual facility for extension to high dimensional models. The optimal strategies which are feedback strategies are obtained by an iterative procedure. The starting strategy for both players is a uniformly mixed strategy over all possibilities. Now the game is played. The players will change their strategies depending on the outcome of the game. If for instance at a certain stage, P1's strategy is (1/3, 1/3, 1/3), i.e. he has three possibilities to choose from and he chooses randomly, and the first of the three strategies has been realized and moreover the outcome of the game is victory for P1, then P1 will change the mixed strategy such that the first strategy will have a greater chance of occurrence (in case he gets into the same situation again). During the second run P1's strategy at that particular stage could for instance be (1/2, 1/4, 1/4). P2 in this case will change his mixed strategies such that this realization of the game with a negative outcome for him, is less likely to occur again. For computational expediency some extra assumptions have been made such as (i) truncation of the state space to a finite set, (ii) a fixed termination time that terminates all paths as draw outcomes beyond the fixed time, (iii) strategies are functions of the current state only and not of the time to go. The problem solved in Falco and Cohen had about two thousand states, maximum number of stages was twenty (players moved alternately) and about fifty thousand runs (plays) were made in arriving as the decomposition of the state space into the following regions: (1) P1 captures P2, (2) P2 captures P1, (3) simultaneous kill (mutual capture), (4) draw.

In the same paper a variation on the model has been considered. The game is then played continuous time and the state space is split up into mutually disjoint cells. The players only know in which cell their opponent is, but do not know their exact position. In this set-up, though not necessarily from the model requirements, it is also assumed that the players can choose from a finite number of actions. The motivation is that a human player will not choose from a continuum of possibilities in practice. The players' strategies, which can be mixed again, only depend on where, in which cell, their opponent is located. The optimal strategies are obtained again by simulation.

This approach is elaborated upon in Kelley and Lefton (1978), who apply a preference-ordering to rational choice between outcomes. In addition to the four mentioned, this paper considers a fifth outcome, vis., (5) one player, say P1, can disengage. A reasonable preference order for P1 is (1), (5), (4), (3), (2) and for P2: (2), (4), (5), (3), (1). The paper (Kelley, Lefton and Cliff, undated) uses the same techniques to describe so-called reprisal strategies in pursuit games. A reprisal strategy is a control policy based upon the extrapolation of the opponent's current departures from optimality. (Note that the optimal or saddlepoint strategies in a zero-sum game give a lower-respective upperbound for one's outcome. If one player constantly uses a non-optimal strategy with a predictable pattern, then the other player can exploit this feature and play better than his saddlepoint strategy.)

Cilleti, Meyer and Salmon (1973) apply the techniques of reachable sets to role determination in air combat. For each fixed time the reachable sets of both airplanes are calculated (i.e. all positions which can be achieved from given initial conditions) and in addition, the bullet-reachable sets. It is possible to include the finite bullet speed in this set up. One airplane, say P2, is vulnerable for a shot down if for a fixed time his reachable set is inside P1's bullet reachable set and simultaneously, P1's reachable set does not intersect with P2's bullet reachable set.

Other approaches to role determination and related matters, specifically with respect to airplane dogfight problems can be found in Kelley (1975a, 1975b) and Kelley and Lefton (1977). In these papers, the description of the aircraft-maneuverability by means of the energy-state model stays central.

IV.2. Worst Case Designs

The idea of replacing uncertainty of unknowns by the worst possible is age old and universal in engineering and military analysis. However, in dynamic setting, the literature is more restricted. One gets a bound on the outcome of the optimal control problem by solving the game in which "Nature" tries to do the worst possible, i.e. it tries to maximize the criterion which we try to minimize. Thus a zero-sum differential game arises. In Gutman and Letimann (1976) the stabilization of vibrations in a two-story building during an earthquake is considered, using a worst case design. See also Leitmann and Gutman (1976). In estimation of the state of dynamic systems, we have the work of Schweppe (1973B), Johansen (1966) and Basar and Mintz (1972).

IV.3. Models of Arms Races and Battles

In Richardson (1960) the following model was developed to describe armament buildups between two countries:

$$\dot{x}_1 = -\alpha x_1 + u_1 \tag{4.1}$$

$$\dot{x}_2 = -\beta x_2 + u_2$$

where

$$u_1 = \sigma x_2 + g \ , \quad u_2 = p x_1 + h \tag{4.2}$$

The scalars x_1 and x_2 are measures of the armament level of the two countries. The underlying assumptions for this model are that for each nation its armament level will tend

(i) to increase proportionally to the armament level of the other nation;

(ii) to decrease proportionally to the economic burden corresponding to the size of its own existing armament;

(iii) to increase guided by its grievance and hatred towards the other nation.

Instead of $u_1(t)$ and $u_2(t)$ to be given by (4.2), we will assume that they are control variables and denote the expenditure on weapons at time t by each nation. In Saaty (1968), thus a differential game has been formulated by adding the cost functional

$$\int (a_1x_1^2 + a_2x_2^2 + \ell_1u_1^2 + \ell_2u_2^2)dt ,$$

where a_1, a_2, ℓ_1 and ℓ_2 are given weights. In a couple of papers by Simaan and Cruz (1975a, 1975b, undated), nonzero-sum versions of this differential game have been studied.

If two nations are engaged in a war, Lanchester developed the following model where x_1 and x_2 have the same meaning as above:

$$\dot{x}_1 = -\beta x_2 , \quad \dot{x}_2 = -\alpha x_1 ,$$

here α is the "hitting power" of the first nation and β the one of the second. Optimal control extensions of the Lanchester model have been considered in Taylor (1978). An up-to-date survey can be found in a companion paper in this volume (Karr, 1983). In Isaacs (1965B), extensions to zero-sum differential games have been studied. One of Isaacs' problems is the following, which he called the war of attrition and attack:

$$\dot{x}_1 = m_1 - \alpha u_2 x_2$$

$$\dot{x}_2 = m_2 - \beta u_1 x_1$$

with $0 \leq u_1, u_2 \leq 1$, and

$$J = \int_0^T \{(1 - u_2)x_2 - (1 - u_1)x_1\}dt .$$

The parameters α and β are constants; m_1 is the rate of capacity to manufacture weapons for nation i.

In Berkowitz and Dresher (1960), the problem of allocating two types of aircraft (bombers and fighters) among three different air tasks (counter air, air defense, and support of ground operations), has been considered as a discrete time differential game. See also Berkowitz (1975).

In Kumar (1980), the problem is considered of a maneuvering evader, attempting to reach a certain destination, while it is being fired upon by the gun (the "pursuer"). The gun has a maximum firing frequency and the travel time of the projectiles fired has been taken into account. The crucial feature of this problem is that both players have to use mixed strategies.

Another related problem is that of Duels. Two players approach each other along a line, each having finite number of bullets and having specified aiming accuracy as a function of distance. The goal is to hit your opponent first. Different versions of this problem under various assumptions on information (such as noisy or silent bullets) have been studied. See Kimeldorf (1983).

IV.4. Blind Pursuit Evasion

Blind pursuit-evasion was initiated by search games in Isaacs (1965), specifically by the Princess and Monster game. In this search game, both the searcher (the Monster) and the hider (the Princess) can move in a bounded region. Each of the players cannot see the other until the distance between them is less than a given quantity, r, and at this very moment capture occurs. Capture time is the criterion, to be minimized by the Monster and to be maximized by the Princess.

The problem where both the searcher and hider can move on the boundary of a circle, has been solved in Alpern (1974), Foreman (1977), and Zelikin (1972). Discrete versions of this problem were considered in Wilson (1972) and Worsham (1974). If the Monster's maximum speed is 1 and the Princess' speed is > 1, then the Princess will also use a speed 1 for her optimal strategy. Under the assumption that $r = 0$ (not very essential), the cohato (coin-half-tour) move is an important element in both players' strategies. The cohato move is the mixed strategy which is the equiprobable selection of the left or right half tour, thus travelling with speed 1 to the point opposite from the starting point. After these points have been reached by the players, the cohato moves are repeated, independent of the previous moves. If prior to the start of a cohato move, one player only knows that the other player's position is uniformly distributed along the circle, and vice versa, then, if during the cohato move the game does not terminate, at the end of this move the position of one player is again uniformly distributed along the circle as viewed through the eyes of his opponent, and vice versa. The capture chance during each cohato move is 1/2 and thus the total capture change increases if the cohato strategy is repeated. The techniques used for the derivation of the saddlepoint strategies for the Princess and Monster game were highly problem oriented and it was difficult if not impossible to extend them to search games in areas of higher dimension. A completely different approach was used in Gall (1979) for searches along arbitrary networks and extensions to blind pursuit-evasion in a two (or higher) dimensional region Q. In the latter case an ε-optimal strategy of the searcher is as follows. Q is covered by a set of rectangles, $Q_1, \ldots, Q_M$. The searcher chooses randomly a rectangle Q_i, goes into it as fast as possible and moves a predetermined time within this rectangle according to a specific strategy, then chooses another rectangle at random, etc. An ε-optimal strategy for the hider is: choose a point $x_1 \in Q$, using a uniform probability distribution in Q, go there and stay there for a predetermined time, then choose another point $x_2 \in Q$ according to the same distribution, etc. The expected capture time is about $\mu/2r$, provided Q and r satisfying some modest assumptions. Here μ is the Lebesgue measure of Q. This is equivalent to as to say that the expected capture time is the time it takes for the searcher to search the whole region Q, i.e. to cover each point of Q at least once. A similar statement for networks--expected capture time is the time it takes to transverse the whole network--cannot be made as has been shown by Fitzgerald (1979).

V. EXTENSIONS TO NONZERO-SUM AND STOCHASTIC GAMES

As we mentioned briefly in the Introduction, beyond the confines of zero sum deterministic difference games, two natural generalizations suggest themselves. The first direction involving relating the assumption of zerosum to nonzero sum; the second, via the introduction of randomness. Ultimately, of course we complete the circle and come back to the most general games in extensive form which was proposed by Von Neumann in the starting point of game theory. In below sections, we shall sketch the developmental framework for NZS and stochastic DG and some of the salient features of such games. Although considerable literature has accumulated over the past decade on these two subjects, applications up to now have been mostly qualitative and potential in nature. This is understandable in view of the difficulties with ZS deterministic DG discussed earlier. We included this treatment of the NZS and stochastic case only to stress our belief on the future directions of research for DG rather than to attempt a comprehensive review of these subjects.

V.1. Nonzero Sum Differential Games

Since DG is merely a very specialized form of general extensive games, all the conceptual issues associated with general games comes into play once we abandoned the zero sum assumption. The analog of the saddle point condition Eq. (2.9) is the Nash equilibrium condition which for the two player case is

$$J_1(\gamma_1^*, \gamma_2^*) \le J_1(\gamma_1, \gamma_2^*) \quad \forall \gamma_1 \in \Gamma_1 \tag{5.1}$$

$$J_2(\gamma_1^*, \gamma_2^*) \le J_2(\gamma_1^*, \gamma_2) \quad \forall \gamma_2 \in \Gamma_2$$

where J_i , $i = 1, 2$ are the cost criteria for the players and γ_1^* , γ_2^* is known as the Nash equilibria strategy pair. It is well known in game theory that as a solution concept, Nash equilibrium solution is much weaker compared to the saddle point solution. The principle difficulty is the fact that two Nash equilibria (γ_1^*, γ_2^*) and (γ_1^0, γ_2^0) are in general not equivalent nor interchangable as they were in the ZS case. Thus while γ_1^* is the best reaction to γ_2^* for player 1, nothing can be said about $J_1(\gamma_1^*, \gamma_2^0)$. In fact, in general, if the players cooperate, then there generally exist $(\gamma_1^{\#}, \gamma_2^{\#})$ such that $J_i^{\#}$ for $i = 1, 2$ are both better than J_i^* . Consequently, other solution concepts both noncooperative and cooperative needs to be investigated. Thus, a fair amount of literature in the early days and currently amounts to a marriage of mainstream game-theoretic concepts with the techniques of control theory and differential games, and the explicit determination of strategies and solutions for nonzero sum games in strategic or characteristic function form. See Starr and Ho (1969), Case (1969), Liu (1973b), Lukes (1971), Ray and Blaquiere (1980), and Gabasov, Gorokhovik and Kirillova (1978).

The proliferation of solution concepts does bring in a richness to differential game models hereto nonexistent in the zero sum case. Many economic problems of competition are more adequately captured as nonzero-sum models; such as advertising (Olsder, 1976; Leitmann and

Schmitendorf, 1978); trade (Case, 1971, 1973, 1979B); negotiation (Liu, 1973a, 1974; Leitmann and Chen, 1980; Leitmann, 1973; Clemhout Leitmann and Wan, 1975, 1976) and economic policies (Leitmann and Wan, 1978, 1979; Pindyck, 1977; Clemhout, Leitmann and Wan, 1971, 1976; Wishart and Olsder, 1980; Ozguner and Perkins, 1979; Simaan and Takayama, 1978).

In principle, the study of NZSDG should involve extensive and interesting research regarding various kinds of known or yet-to-be discovered singular surfaces. However, to date relatively little is known along such lines ((Sarma and Prasad, 1976) is a paper in this direction). Partly this can be explained by the lack of researchers with a combination of serious commitment to both economies and differential games. Unlike the situation with various pursuit-evasion DGS, there is a fair amount of vertical structure in both fields to be acquired before one can credibly work in this area. The most extensive collection of models can be found in Case (1979B).

The sequential nature of a DG introduces another feature which is often suppressed in the strategic form of a game. Because each player makes a series of decisions, and because in deterministic situations past decision of the players can be inferred from past state history and made known to players currently, the possibility of a generalized "threat" (or incentive) exist. The basic idea is the one used in resolving the Prisoner's Dilemma Problem in repeated games. One player can adopt a strategy which essentially says to the other player that unless you choose such and such decisions, I shall take such and such punitive actions. By such a device, one can create equilibrium solutions almost at will. See Basar (1976, 1977b, 1979), Basar and Selbuz (1979), Papavasoloupos and Cruz (1979), Tolwinski (1980), Basar and Olsder (1979). This phenomenon does not occur in ZS case since saddle point solution cannot be improved upon by threat due to equivalence and interchangability.

In games where one player's decision is dependent on that of the other is commonly referred to as a Stackelberg problem. In the dynamic setting, see Simaan and Cruz (1973a, 1973b), Cruz (1978, 1975), and Olsder (1977a).

An extreme version of nonzero sum game is the case where $J_1 = J_2$. In this case we have the special class of so-called team problems in contrast with the purely conflict situation of the zero sum case. Deterministic team problems are merely simple constrained optimization problems. New concepts arise only when each player has access to different information. Such issues are best left to the discussion in connection with the stochastic case. We mention a separate survey article for those who are interested (Ho, 1980S).

Optimal control theory with vector payoff is another way of looking at multiperson nonzero sum cooperative differential games. The emphasis here is not so much the study of coalition structure as in main-stream game theory but rather the characterization of the pareto-optimal or noninferior set.[3] A comprehensive survey by Stadler (1979) exists. A more abstract solution concept exists, involving prefhrence and comparison relations, from which both Nash and Pareto (and also other equilibrium-definitions) follow as special cases (Blaquiere, 1976a).

V.2. Information Structure and Stochastic DG

One major by-product of the research on DG and extensive games is a deeper appreciation of the intricacies of the information structure of a game, i.e. "who knows what when?" and its corollary "who should know what when?" For one thing, the size of the class of admissible strategies is directly related to the finess of the information structure.[4] Not only different admissible strategy classes produce different payoffs for the same problem but they may possess widely different qualitative properties. Take the example of a simple pursuit-evasion game in two dimensions. Whether or not the pursuer-evader pair plays the game based on initial relative position (open loop) or current relative position (closed loop) makes a big difference in payoff and strategies employed if the playing time is sufficiently long. Such phenomenon does not occur in one person optimization problems. Of course open vs. closed loop are merely the limits of the more general noisy observation situation where the noise variances are neither infinite nor zero. In such cases, we have stochastic differential games. Stochastic DG, however, are not easy extensions of one person stochastic control theory because of one additional complication; namely, different players may have access to different information.[5] This natural and seemingly simple assumption creates a host of new conceptual problems which is still not completely resolved or widely understood today. The basic difficulty has to do with the fact that one person's current information may be dependent on another person's strategy at earlier times. Consequently, the following vicious cycle develops as illustrated in the Figure below:

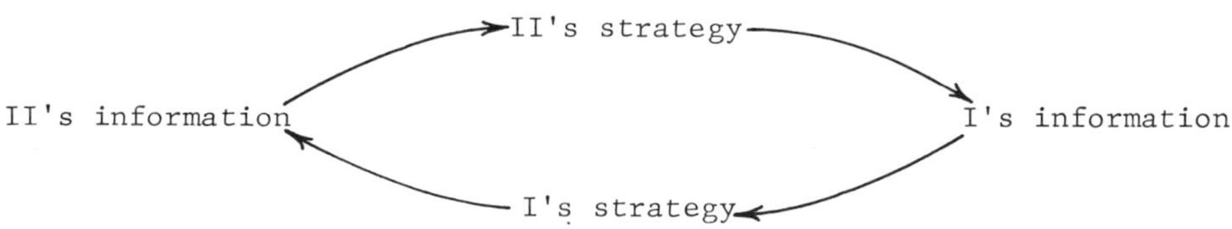

A detailed and tutorial treatment of this basic problem can be found in Ho (1980S). We mention some particular ramifications in association with DG.

(i) The second-guessing problem. This is the familiar problem of "I know that you know that I know that you know...." In particular, for linear-quadratic-Gaussian stochastic DG with finite dimensional dynamic systems, we have the result that the saddle point strategies are not finite dimensional (Willman, 1969; Bagchi and Olsder, 1980). If one postulates a finite dimensional strategy for one player, then he finds that the reaction of the other player involves a strategy with higher dimensions. This iterated reaction can only reach equilibrium when the strategies become infinite dimensional or by restricting the admissible strategy class by assumption, see Rhodes and Luenberger, 1969).

(ii) Non Optimality of the Control-Dependent Kalman Filter. Suppose the pursuer builds a Kalman filter for an evading dynamic system using partial state observation $z_p(t)$ and measurement of the evader's control $u_e(t)$ in the standard manner. The estimate produced by this Kalman Filter is not optimal. This is because $u_e(t)$ is in general a function of evader's own observation $z_e(t)$ and

$$\hat{x}_e(t) \equiv E(x_e(t)/z_p[t_0, t], u_e[t_0, t])$$

$$\equiv E(x_e(t)/z_p[t_0, t], \gamma_e[z_e[t_0, t]])$$

where $z[a,b]$ stands for the measurement history $z(\tau)$ $z \le \tau \le b$. In the one person case, or in the case where z_p include $\underline{z}_e$, we have

$$\hat{x}_e(t) = E(x_e(t)/z_p[t_0, t])$$

and no new information is conveyed by $u_e(t)$ to the pursuer. Consequently, LQG stochastic DG are far easier to solve in the case when $z_p(t) \supseteq z_e(t)$. See Behn and Ho (1968), Ho (1974), Balakrishnan (1976B), Elliott (1977b). Again one may avoid this problem by assumption, i.e. restricting the use of $u_e(t)$ in the strategy of pursuer, see Casalino *et al*. (1980).

(iii) Mixed strategy equilibrium. Primarily because of the continuous nature of the variables involved in a DG, most saddle points of DG are in pure strategies. Genuine mixed strategy solutions occur in minimax estimation problems treated by Basar and Mintz (1972).

Another direction in differential games, though not restricted to DG's, is related to imperfect information on the system with no statistical description on the possible errors. The state of the dynamic system is described by means of certain informational sets in the state space that correspond to all admissible states of the system compatible with the available information. Some references in this direction are Chigir (1977), Kurshanskii (1978), Krasovskii (1975) and Schweppe (1973B).

Value of Information. While it is reasonably intuitive that private information available to one player can never hurt (Witsenhausen, 1971) his payoff, it is not at all obvious what effect public information has on the payoff of a stochastic DG. The problem is complicated by the fact that explicit determination of the payoff in terms of the parameters of the DG problem is generally difficult. Furthermore, the definition of the values of a piece of public information is no longer clear cut. This question has been addressed generally in Sun and Ho (1976) and Witsenhausen (1977). For the case of LQG DG, we have the intuitively satisfying result that the new public information benefits the more manueverable player more (Sun and Ho, 1976b, 1976c), e.g. consider a pursuit-evasion game in dense or light fog.

In the one person case, the value of information question arises because information must often be acquired at a cost. In the two person case, additional trade off considerations enter because information sometimes can only be acquired publicly. (See Olsder (1977b) for the case where information can be acquired independently of each other.) Thus, there is an indirect cost since one's opponent can also take advantage of such public information. Examples of such problems are many. One involves communication among team members in a game over a nonsecure channel (Ho and Sun, 1974). Another problem involves a pursuer guarding a coastline. P can improve his chance capture by using a search light which unfortunately also reveals his position to the evader. See Hexner (1979) for a similar problem.

Stochastic Markov Chain Games. When both the state and control are discrete, we have the game analog of the Markov chain decision problems (Sobel, 1971; Parthasarathy and Stern, 1977). The seminal paper by Shapley (1953) set the style for almost all work in this area. Beyond the simplicity of the Markov Chain, there is little structure to the problem. Consequently, research in this area tends to focus on existence and theoretical questions. Real world applications are rare (Shubik and Sobel, 1980; Eaton and Zadeh, 1962; Kushner and Chamberlain, 1969). This is further complicated by computational difficulties since the number of states are combinatorily large in real problems.

CONCLUSION

It is somewhat difficult to assess the total contribution of differential games. From a practical point of view, one cannot claim as many success stories as its one player counterpart (control theory) can do. Whatever applications there are, they have been mostly qualitative. For theorists, the subject did provide many hard technical problems of analysis. While these tasks are necessary and legitimate as part of mathematics, they did not help much in solving practical pursuit-evasion problems which first motivated the study. It seems to the authors that the main contribution of differential games lies not so much in the specific solution of real world problems, but in the way we look at these problems. Differential games provided a meeting ground for control theory and game theory. It brought about a resurgence of interest in games in extensive form which mainstream game theory tends to ignore. Considerable new insights and concepts were developed which were not part of either regular control or game theory. This is particularly true for nonzero sum and stochastic differential games. Thus, the role of differential games is somewhat akin to that of a catalyst. By studying a relatively narrowly defined class of problems (Deterministic DG) we are led to a broader view of many-person optimization and rapproachment of decision and control methodology with game theory.

FOOTNOTES

[1]There does exist in English language a survey by Zauberman (1975B). Readers, however, should be warned that this is not a book in the usual sense but rather a collection of more or less unorganized personal notes on various USSR works in this area.

[2]Other names used are: P1 and P2, I and II, u- and v-player, etc.

[3]The set of payoffs which is not dominated by at least one another set of payoffs in which each player is better off.

[4]Since strategy is a map from information to decision the more finely detailed the information structure the more sophisticated can be the decision. For example, consider the strategies "honesty is the best policy" vs. "honesty is the best policy except when a little white lie may protect the feelings of a loved one." The latter strategy requires more information to implement.

[5]Note that in simple open and closed loop cases, one can still regard each player as having the same information.

REFERENCES

Books on Differential Games and Books Referenced

[1] Balakrishnan, A. V., Applied Functional Analysis (Springer Verlag, New York & Vienna, 1976).

[2] Basar, T. and Olsder, G. J., Dynamic Noncooperative Game Theory (Academic Press, New York, 1982).

[3] Bernhard, P., Commande Optimale, Decentralisation et Jeux Dynamiques (Dunod, Paris, 1976).

[4] Blaquiere, A., Gerard, F., and Leitmann, G., Quantitative and Qualitative Games (Academic Press, New York, 1969).

[5] Blaquiere, A. and Leitmann, G., Jeux Quantitatifs (Gauthier-Villars, Paris, 1969).

[6] Bryson, A. E. and Ho, Y. C., Applied Optimal Control (Hemisphere, Washington, D. C., 1975).

[7] Case, J. H., Economics and the Competitive Process (New York University Press, New York, 1979).

[8] Danskin, J., Theory of Max-Min and Its Application to Weapons Allocation Problems (Springer Verlag, New York & Vienna, 1967).

[9] Friedman, A., Differential Games (Wiley, New York, 1971).

[10] Friedman, J. W., Oligopoly and the Theory of Games (North Holland, New York, 1977).

[11] Hajek, O., Pursuit Games (Academic Press, New York, 1975).

[12] Isaacs, R., Differential Games (first edition, Wiley, New York, 1965; second edition, Kruger Publishing Company, Huntington, New York, 1975).

[13] Krassovski, N. and Soubbotine, A., Jeux Differentials (French translation, original text published in 1974; Editions Mir, Moscow, 1977).

[14] Leitmann, G., Cooperative and Non-cooperative Many Player Differential Games, CISM Monograph No. 190 (Springer-Verlag, Vienna, 1974).

[15] Owen, J., Game Theory (Saunders, Philadelphia, 1968).

[16] Richardson, L. F., Arms and Insecurity (Boxwood, Pittsburgh, 1960).

[17] Saaty, T. L., Mathematical Models of Arms Control and Disarmament (Wiley, New York, 1968).

[18] Schweppe, F., Uncertain Dynamic Systems (Prentice Hall, Englewood Cliffs, New Jersey, 1973).

[19] Telser, L. G., Competition, Collusion, and Game Theory (Aldine, Atherton, Chicago and New York, 1972).

[20] Vorob'ev, N. N., Game Theory (Springer Verlag, Vienna, 1977).

[21] Zauberman, A., Differential Games and Other Game-Theoretic Topics in Soviet Literature: a survey (New York University Press, New York, 1975).

Other Survey Papers and Bibliographical Sources, Edited Books

[22] Bernhard, P. (ed.), Seminaire des 21-25 Juin Les Jeux Differentiels, Center d'Automatique de l'Ecole National des Mines de Paris (1971).

[23] Blaquiere, A. (ed.), Topics in Differential Games (North Holland, Amsterdam, 1973).

[24] Clemhout, S. and Wan, H. Y., Interactive economic dynamics and differential games, Journal of Optimization Theory and Applications 27:1 (1979) 7-30.

[25] Grote, J. D. (ed.), The Theory and Application of Differential Games (Reidel, Hingham, Massachusetts, 1975).

[26] Hagedorn, P., Knobloch, H. W., and Olsder, G. J. (eds.), Differential Games and Applications, Lecture Note No. 3 in Control and Information Sciences (Springer Verlag, Vienna, 1977).

[27] Ho, Y. C. (ed.), Proceedings of the First International Conference on the Theory and Applications of Differential Games (1969).

[28] Ho, Y. C., Team decision theory and information structures, Proceedings of IEEE (June 1980).

[29] Ho, Y. C. and Mitter, S. K. (eds.), Directions in Large Scale Systems (Plenum, New York, 1976).

[30] Leitmann, G., and Marzollo, A. (eds.), Multicriteria Decision Making, CISM Monograph No. 211 (Springer Verlag, Vienna, 1975).

[31] Leitmann, G. (ed.), Multicriteria Decision Making and Differential Games (Plenum, New York, 1976).

[32] Liu, P. T. (ed.), Dynamic Optimization in Mathematical Economics (Plenum, New York, 1980).

[33] Liu, P. T. and Roxin, E. O. (eds.), Differential Games and Control Theory III (Dekker, New York, 1979).

[34] Roxin, E. O., Liu, P. T., and Sternberg, R. L. (eds.), Differential Games and Control Theory I (Dekker, New York, 1974).

[35] Roxin, E. O., Liu, P. T., and Sternberg, R. L. (eds.), Differential Games and Control Theory II (Dekker, New York, 1976).

[36] Tinjanskii, N. T. and Zhukovskii, V. I., Nonzero-sum differential games, in: Gamkrelidse, R. V. (ed.), Mathematical Analysis, Vol. 15 (Moscow, 1977, in Russian).

[37] Tsokos, C. P. and Nicholas, W. G., On some stochastic differential games, in: Bharucha-Reid, A. T. (ed.), Probabilistic Methods in Applied Mathematics (Academic Press, New York, 1973).

References Cited

[38] Al-brekht, E. G. and Loginov, M. I., On the continuous dependence of a linear encounter game on a parameter, Journal of Applied Mathematics and Mechanics (translated of PMM), 40 (1976) 188.

[39] Alpern, S., The search game with mobile hider on the circle, in: Roxin, E. O. et al. (eds.), Differential Games and Control Theory (Dekker, New York, 1974).

[40] Anderson, G. M., Feedback control for a pursuing spacecraft using differential dynamic programming, AIAA Journal 13:8 (1977) 1084-1088.

[41] Anderson, G. M. and Bohn, G. D., A near-optimal control law for pursuit-evasion problems between two spacecraft, AIAA Journal 13:8 (1977) 1203-1205.

[42] Anderson, G. M. and Grazier, V. M., Barrier in pursuit-evasion problems between two low-thrust orbital spacecraft, AIAA Journal 14:2 (February 1975).

[43] Bagchi, A. and Olsder, G. J., Linear stochastic pursuit evasion games, Journal of Applied Mathematics and Optimization (1980).

[44] Baron, S., Ho, Y. C., and Kleinman, D. L., A new approach to aerial combat games, NASA Report CR-1626 (February 1970).

[45] Basar, T., On the uniqueness of the Nash solution in linear-quadratic differential games, International Journal of Game Theory 5 (1976) 65-90.

[46] Basar, T., Two general properties of the saddle-point solutions of dynamic games, IEEE Transactions of Automatic Control (February, 1977). (a)

[47] Basar, T., Informationally nonunique equilibrium solutions in differential games, SIAM Journal on Control and Optimization, 15 (1977) 636-660. (b)

[48] Basar, T., Information structures and equilibria in dynamic games, in: New Trends in Dynamic System Theory and Economics (Academic Press, New York, 1979), pp. 3-55.

[49] Basar, T. and Mintz, M., Minimax terminal state estimation for linear plants with unknown forcing functions, International Journal of Control 16:1 (1972) 49-70.

[50] Basar, T. and Olsder, G. J., Team-optimal closed-loop Stackelberg strategies in linear-quadratic differential games, Proceedings 17th Allerton Conference on Communication, Control and Computing (Urbana, Illinois, 1979).

[51] Basar, T. and Selbuz, H., Closed-loop Stackelberg strategies with applications in the optimal control of multilevel systems, IEEE Transactions on Automatic Control, AC-24 (1979) 166-179.

[52] Behn, R. D. and Ho. Y. C., On a class of linear stochastic differential games, IEEE Transactions on Automatic Control, AC-13 (1968) 227-239.

[53] Bensoussan, A., in: Kuhn and Szego (eds.), Differential Games and Related Topics (North Holland, Amsterdam, 1968).

[54] Berkovitz, L. D., The tactical air game: a multimove game with mixed strategy solution, in: Grote, J. D. (ed.), The Theory and Applications of Differential Games (Reidel, Hingham, Massachusetts, 1975) pp. 169-178.

[55] Berkovitz, L. D. and Dresher, M., Allocation of two types of aircraft in tactical air war: a game-theoretic analysis, Operations Research 8 (1960) 694-706.

[56] Bernhard, P., Linear pursuit-evasion games and the isotropic rocket, Sudaar No. 413, Stanford University, Center for Systems Research (1970).

[57] Bernhard, P., Condition de coin pour les jeux differentiels, Seminaire des 21-25 Juin, Les Jeux Differentiels, Centre d'Automatique de l'Ecole Nationale Superieure des Mines de Paris (1971).

[58] Bernhard, P., Sur la commandabilite des systemes lineares discrets a deux joueurs, Revue R.A.I.R.O. no. Aout 1972-J-3 (1972) 53-68.

[59] Bernhard, P., New results about corners in differential games, including constraints, 6th Triennial IFAC World Congress, Boston (1975).

[60] Bernhard, P., Singular surfaces in differential games and applications, in: Hagedorn, P., Knobloch, H. W. and Olsder, G. J. (eds.), Differential Games and Applications, Lecture Note No. 3 in Control and Information Sciences (Springer Verlag, Vienna, 1977).

[61] Bernhard, P., Linear-quadratic, two-person, zero-sum differential games: necessary and sufficiency conditions, Journal of Optimization Theory and Applications 27:1 (1979) 51-69.

[62] Bernhard, P., Principe d'optimalite et information necessaire pour la conception dans le cas le plus defavorable, Cahiers de Mathematiques de la Decision, No. 7923, Universite de Paris-Dauphine (undated).

[70] Blaquiere, A., An introduction to differential games, in: Kuhn, H. W. and Szego, G. P. (eds.), Differential Games and Related Topics (North Holland, Amsterdam, 1971) pp. 47-81.

[71] Blaquiere, Austin, Differential games with time lag, in: Techniques of Optimization (Academic Press, New York, 1972) pp. 287-297.

[72] Blaquiere, A., Une generalisation du concept d'optimalite et de certaines notions geometriques qui s'y rattachent, Institut des Hautes Etudes de Belgique, Cahiers du Centre d'Etudes de Recherche Operationnelle, Vol. 18, No. 1-2, Bruxelles (1976), pp. 49-61. (a)

[73] Blaquiere, A., Dynamic games with coalitions and diplomacies, in: Ho, Y. C. and Mitter, S. K. (eds.), Directions in Large-Scale Systems (Plenum, New York, 1976), pp. 95-115. (b)

[74] Blaquiere, A., Differential games with piece-wise continuous trajectories, in: Hagedorn, P., Knobloch, H. W. and Olsder, G. J. (eds)., Differential Games and Applications, Lecture Note No. 3 in Control and Information Sciences (Springer Verlag, Vienna, 1977).

[75] Blaquiere, A., Wave mechanics as a two player game, in: Blaquiere, A. and Fer, F. (eds.), Proceedings of the International Seminar on Mathematical Theory of Dynamical Systems and Microphysics, CISM Course and Lectures (Springer Verlag, Vienna, 1980).

[76] Blaquiere, A. and Caussin, P., Differential games with time lag, in: Blaquiere (ed.), Topics in Differential Games (North-Holland, Amsterdam, 1973) pp. 151-178.

[77] Blaquiere, Austin and Gerard, Francoise, On the geometry of optimal strategies in two-person games of kind, Journal of Computer and Systems Sciences 2:3 (1968) 288-304.

[78] Breakwell, J. V., Examples elementaires, seminaire des 21-25 Juin, Les Jeux Differentels, Centre d'Automatique de l'Ecole Nationale Superieure des Mines de Paris (1971).

[79] Breakwell, J. V., Some differential games with interesting discontinuities, Internal Report, Stanford University (1973).

[80] Breakwell, J. V., Pursuit of a faster evader, in: Grote, J.D. (ed.), Theory and Applications of Differential Games (Reidel, Hingham, Massachusetts, 1975) pp. 243-256.

[81] Breakwell, J. V., Zero-sum differential games with terminal payoff, in: Hagedorn, P. et al. (eds.), Differential Games and Applications (Springer Verlag, Vienna, 1977).

[82] Breakwell, J. V., Delayed decisions in zero-sum differential games, Internal Report, Stanford University (undated).

[83] Breakwell, J. V. and Lewin, J., The surveillance evasion game of degree, Journal of Optimization Theory and Applications 16 (1975).

[84] Casalino, G., Davoli, F., and Zoppoli, R., Instantaneous exchange of imperfect messages within a delayed sharing pattern information structure, Large Scale Systems (1980).

[85] Case, J. H., Towards a theory of many players differential games, SIAM Journal of Control 7:2 (1969) 179-197.

[86] Case, J., On Ricardo's problem, Journal of Economic Theory 3:2 (1971) 134.

[87] Case, J., Differential trading games, in: Blaquiere, A. (ed.), Topics in Differential Games (North Holland, Amsterdam, 1973).

[88] Chernous'ko, F. L., A problem of evasion from many pursuers, Journal of Applied Mathematics and Mechanics (translation of PMM) 40 (1976) 950

[89] Chigir, S. A., The game problem on the dolichobrachistochrone, PMM 30:6 (1976) 1003-1013.

[90] Chigir, S. A., On an alternative in a differential evasion game with incomplete information, Journal of Applied Mathematics Mechanics (translation of PMM) 41 (1977) 1.

[91] Chikrii, A. A., The evasion problem in nonstationary differential games, PMM 39:5 (1975) 780-787.

[92] Chikrii, A. A. and Rappoport, I. S., Linear Problem of Pursuit by Several Guided Objects, translated from Kibernetika 3 (1978) 86-92 (Plenum, New York, 1977) pp. 407-414.

[93] Ciletti, M. D., Results in the theory of linear differential games with an information time lag, Journal of Optimization Theory and Applications 5:4 (1970) 347-362. (a)

[94] Ciletti, M. D., On the contradiction of bang-bang-bang surfaces in differential games, Journal of Optimization Theory and Applications 5:3 (1970) 163-169. (b)

[95] Ciletti, M. D., Meyer, L., and Salmon, D. M., Application of reachable sets techniques to air combat analysis, Paper No. 73-232 of the AIAA 11th Aerospace Science Meeting, Washington,1973.

[96] Clemhout, S., Leitmann, George, and Wan, H. Y., Jr., A differential game model of oligopoly, Journal of Cybernetics 3:24 (1973).

[97] Clemhout, S., Leitmann, G., and Wan, H. Y., Jr., Bargaining under strike: a differential game view, Journal of Economic Theory 11:1 (August 1975) 55-67.

[98] Clemhout, S., Leitmann, George, and Wan, H. Y., Jr., Equilibrium patterns for bargaining under strike: a differential game model, in: Proceedings Conference on Directions in Decentralized Control, Many-Person Optimization and Large Scale Systems (Plenum, New York, 1976), pp. 207-234.

[99] Cockayne, E., Plane pursuit with curvature constraints, SIAM Journal of Applied Mathematics 15:6 (1967) 1511-1516.

[100] Cockayne, E. J. and Hall, G. W. C., Plane motion of a particle subject to curvature constraints, SIAM Journal of Control 13:1 (January 1975).

[101] Cruz, J. B., Survey of Nash and Stackelberg equilibrium strategies in dynamic games, Annals of Economic and Social Measurement 4:2 (1975) 339-344.

[102] Cruz, Jose B., Leader-follower strategies for multilevel systems, IEEE Transactions on Automatic Control, AC-23 (April 1978) 244-255.

[103] Danskin, J., Value in differential games, unpublished manuscript (1973).

[104] Eaton, J. and Zadeh, L., Optimal pursuit strategies in discrete state probabilistic systems, Transactions ASME Journal of Basic Engineering (March 1962) 23-29.

[105] Elliott, R. J., The existence of value in differential games of pursuit and evation, Journal of Differential Equations 12

(1972) 504-523.

[106] Elliott, R. J., Feedback strategies in deterministic differential games, in: Hagedorn, et al. (eds.), Differential Games and Applications, Lecture Note No. 3 in Control and Information Sciences (Springer Verlag, Vienna, 1977) pp. 136-142. (a)

[107] Elliott, R. J., The existence of optimal strategies and saddle-points in stochastic differential games, in: Hagedorn, et al. (eds.), Differential Games and Applications, Lecture Note No. 3 in Control and Information Sciences (Springer Verlag, Vienna, 1977). (b)

[108] Falco, M. and Cohen, U., Strategy synthesis in aerial dogfight game models, Paper No. 73-233 of the AIAA 11th Aerospace Science Meeting, Washington (1973).

[109] Fitzgerald, C. H., The princess and monster differential game, SIAM Journal on Control and Optimization 17:6 (1979) 700-712.

[110] Fleming, W. K., The convergence problem for differential games, Math. Analysis and Applications 3 (1961) 102-116.

[111] Fleming, W. H., The convergence problem for differential games II, Annals of Math. Study 52 (1964) 195-210.

[112] Flynn, J., Pursuit in the circle: lion versus man, in: Roxin, E. O. et al. (eds.), Differential Games and Control Theory (Dekker, New York, 1974) pp. 99-124.

[113] Foley, M. H. and Schmitendorf, W. E., A class of differential games with two pursuers versus one evader, IEEE Transactions on Automatic Control (June 1974).

[114] Foreman, J. G., The princess and the monster on the circle, SIAM Journal of Control and Optimization 15 (1977) 841-856.

[115] Gabasov, R., Gorokhovik, V. V., and Kirillova, F. M., Multi-criterion optimization problems and optimal control theory, in: Reprints 7th IFAC Triennial World Conference (Helsinki, Pergamon Press, 1978), p. 1001.

[116] Gal, S., Search games with mobile and immobile hider, SIAM Journal of Control and Optimization 17:1 (1979) 99-122.

[117] Gamkrelidze, R. V., and Kharatishvili, G. L., A differential game of evasion with nonlinear control, SIAM Journal of Control 12:2 (1974) 332-349.

[118] Getz, W. M. and Leitmann, G., Qualitative differential games with two targets, Journal of Mathematical Analysis and Applications 68:2 (April 1979) 421-430.

[119] Getz, W. M. and Pachter, M., Two-target pursuit-evasion differential games in the plane, Journal of Optimization Theory and Applications (1980). (a)

[120] Getz, W. M. and Pachter, M., Capturability in a two-target "Game of Two Cars," submitted for publication (1980). (b)

[121] Gutman, S. and Leitmann, G., Stabilizing feedback control for dynamical systems with bounded uncertainty, Proceedings of IEEE Conference on Decision and Control, Miami (1976), 94-99.

[122] Gusiantnikov, P. B., On a problem of l-escape, PMM 40:1 (1976) 25-37.

[123] Hagedorn, P. and Breakwell, J. V., A differential game with two pursuers and one evader, Journal of Optimization Theory and Applications 18 (1976) 15-29.

[124] Hagedorn, P. and Carmele, W., A problem in collision avoidance, in: Reprints of the 7th Triennial World Congress of IFAC, Helsinki (Pergamon Press, 1978), pp. 989-995.

[125] Hajek, O., Discontinuous differential equations, I and II, Journal of Differential Equations 32:2 (1979) 149-170, 171-185.

[126] Heymann, M. and Pachter, M., On the max-min pursuit game, Journal of Mathematical Analysis and Applications 70:2 (1979).

[127] Heymann, M., Pachter, M., and Stern, R. J., Max-min control problems; a system theoretic approach, IEEE Transactions on Automatic Control, AC-21:4 (1976). (a)

[128] Heymann, M., Pachter, M., and Stern, R. J., Weak and strong max-min controllability, IEEE Transactions on Automatic Control AC-21:4 (August 1976). (b)

[129] Hexner, G., A differential game of incomplete information, Journal of Optimization Theory and Applications 28:2 (1979).

[130] Ho, Y. C., On the minimax principle and zero-sum stochastic differential games, Journal of Optimization Theory and Applications 13:2 (1974) 343-361.

[131] Ho, Y. C., Bryson, A. E., and Baron, S., Differential games and optimal pursuit-evasion strategies, IEEE Transactions on Automatic Control AC-10:4 (October 1965).

[132] Ho, Y. C. and Chu, K. C., Information structure in many person optimization problems, Automatica 10 (1974) 341-351.

[133] Ho, Y. C. and Sun, F. K., Value of information in two-team zero-sum problems, Journal of Optimization Theory and Applications 14 (1974) 557-571.

[134] Isaacs, R., Differential games, their scope, nature and future, Journal of Optimization Theory and Applications 3:5 (1969).

[135] Jacobson, D. H., On values and strategies for infinite-time linear quadratic games, IEEE Transactions on Automatic Control AC-22:2 (1977) 490-491.

[136] Jarmark, B. S. A., Convergence control differential dynamic programming applied to air-to-air combat, AIAA Journal 14:1 (January 1976).

[137] Johansen, D., Solution of a linear mean square estimation problem when process statistics are undefined, IEEE Transactions

on Automatic Control AC-11:1 (1966) 20.

[138] Karr, A., Lanchester attrition processes and theater-level combat models, these Proceedings.

[139] Kelley, H. J., Some aspects of two-on-one pursuit/evasion, Automatica 9 (1973) 403-404.

[140] Kelley, Henry J., Differential-turning optimality criteria, Journal of Aircraft 12:1 (January 1975) 41-44. (a)

[141] Kelley, Henry J., Differential-turning tactics, Journal of Aircraft 12:12 (December 1975) 930-935. (b)

[142] Kelley, Henry J. and Lefton, Leon, Calculation of Differential-turning barrier surfaces, Journal of Spacecraft and Rockets 14:2 (February 1977) 87-95.

[143] Kelley, Henry J. and Lefton, Leon, A preference-ordered discrete-gaming approach to air-combat analysis, IEEE Transactions on Automatic Control AC-23:4 (August 1978) 652-645.

[144] Kelley, Henry J., Lefton, Leon, and Cliff, Eugene M., Reprisal strategies in pursuit games, JACC, Denver, Colorado (1979), pp. 908-911.

[145] Kimeldorf, G., Duels--an overview, these Proceedings.

[146] Kleimenov, A. F., Problems of conflict control, Journal of Applied Mathematics and Mechanics (translation of PMM) 39 (1975) 212.

[147] Korneev, V. A. and Melrikiran, A. A., Differential game with impulsive control for one of the players, Journal of Applied Mathematics and Mechanics (translation of PMM) 41 (1977) 345

[148] Krasovskii, N. N., On control under incomplete information, Journal of Applied Mathematics and Mechanics (translation of PMM) 40 (1976) 179.

[149] Kumar, P. R., Optimal mixed strategies in dynamic games, IEEE Transactions on Automatic Control (1980).

[150] Kurshanskii, A. B. and Gusev, M. I., Multicriterial game-theoretic problems of control for systems with incomplete information, in: Preprints of 7th IFAC Triennial World Congress, Helsinki (Pergamon Press, 1978), p. 1041.

[151] Kushner, H. and Chamberlain, S., Finite state stochastic games: existence theorem and computation procedures, IEEE Transactions on Automatic Control AC-14:3 (June 1969) 248-254.

[152] Leitmann, George, Collective bargaining: a differential game, Journal of Optimization Theory and Applications 11:4 (1973) 405.

[153] Leitmann, George, On some consequences of playability in differential games, Proceedings of IEEE Conference on Decision and Control, Phoenix (1974), pp. 346-348.

[154] Leitmann, George and Chen, S. F. H., Labor-management bargaining modeled as a dynamic game, Optimal Control Applications and

Methods (1980).

[155] Leitmann, George and Gutman, S., Stabilizing control for linear systems with bounded parameter and input uncertainty, Proceedings 7th IFIP Conference on Optimization Techniques, Springer Verlag, Berlin (1976), pp. 729-755.

[156] Leitmann, G. and Schmitendorf, W. E., Profit maximization through advertising: a nonzero sum differential game approach, IEEE Transactions on Automatic Control AC-23:4 (August 1978) 645-650.

[157] Leitmann, George and Wang, G., Necessary and sufficient conditions for multistage two-person zero-sum games, Journal of Optimization Theory and Applications 4:3 (1969) 145.

[158] Leitmann, George and Wan, Henry Y., Jr., A stabilization policy for an economy with some unknown characteristics, Journal of the Franklin Institute 306:1 (July 1978) 23-33.

[159] Leitmann, G. and Wan, H. Y., Jr., Macro-economic stabilization policy for an uncertain dynamic economy, in: New Trends in Dynamic System Theory and Economics (Academic Press, New Jersey, 1979).

[160] Levine, J., Two person zero-sum differential games with incomplete information--a Bayesian model, in: Roxin and Liu (eds.), Differential Games and Control Theory III (Dekker, New York, 1979).

[161] Lewin, J., The bang-bang-bang problem revisited, Journal of Optimization Theory and Applications 18:3 (1976) 429-432

[162] Lewin, J. and Olsder, G. J., Conic surveillance avoidance, Journal of Optimization Theory and Applications 27:1 (1979) 107-125.

[163] Liu, P. T., Non-zero sum differential games with bargaining solutions, Journal of Optimization Theory and Applications 11:3 (1973). (a)

[164] Liu, P. T., Optimal threat strategies in differential games, Journal of Math. Anal. and Appl. 43:2 (1973). (b)

[165] Liu, P. T., A differential game model for labor-management negotiations, Journal of Optimization Theory and Applications 13:4 (April 1974).

[166] Liu, P. T., A pursuit evasive game for surveillance in convergence zone, Technical Report, Naval Underwater Systems Center, Newport, Rhode Island (February 1977).

[167] Lukes, D. L., A global theory for linear-quadratic differential games, Journal of Math. Analysis and Application 33 (1971) 96-123.

[168] Mageirou, E., Value and strategies for infinite duration linear quadratic games, IEEE Transactions on Automatic Control 21 (1976) 547-551.

[169] Maksimov, V. I., Linear Diff. Encounter Game with a Functional Journal of Applied Mathematics and Mechanics (translation of

PMM) 40 (1976) 301. (a)

[170] Maksimov, V. I., An alternative in a differential-difference game of approach-evasion with a functional target, Journal of Applied Mathematics and Mechanics (translation of PMM) 40 (1976) 936. (b)

[171] Masle, J. F., Problemes quantitatif et qualitatif lies en jeux differentiels, Thesis, University IX, Paris, Dauphine UER Mathematiques de la Decision (1976).

[172] Merz, A. W., The homicidal chauffeur--a differential game, Sudaar No. 418, Guidance and Control Laboratory, Stanford University (1971).

[173] Merz, A. W., The game of two identical cars, Journal of Optimization Theory and Applications, 9:5 (May 1972).

[174] Merz, A., Optimal evasive maneuvers in maritime collision avoidance, Navigation 20:2 (1973) 144-152.

[175] Merz, A. W., Application of differential game theory to role determination in aerial combat, NASA Cr-137713 (1975).

[176] Merz, A. W., A different game solution to the coplanar tail-chase aerial combat problem, NASA Cr-137809 (1976).

[177] Miloh, T. and Sharma, S. D., Maritime collision avoidance as a differential game, Bericht Nr. 329, Institute für Schiffbau der Universtät Hamburg (1976).

[178] Miloh, T. and Sharma, S. D., Critical maneuver for avoiding collision at sea, Internal Report, Institut für Schiffbau der Universtaet Hamburg (undated).

[179] Olsder, G. J., Some thoughts about simple advertising models as differential games and the structure of coalitions, in: Ho, Y. C. and Mitter, S. K. (eds.), Directions in Large-Scale Systems (Plenum, New York, 1976).

[180] Olsder, G. J., Information structures in differential games, in: Sternberg, R. L. (ed.), Proceedings of the 2nd Kingston Conference on Differential Games and Control Theory (Dekker, New York, 1977). (a)

[181] Olsder, G. J., On observation costs and information structures in stochastic differential games, in: Hagedorn, P. et al. (eds.), Proceedings of a Workshop on Differential Games and Applications, Lecture Note No. 3 in Control and Information Sciences (Springer Verlag, Vienna, 1977). (b)

[182] Olsder, G. J. and Breakwell, J. V., Role determination in a aerial dogfight, International Journal of Game Theory 3 (1974) 47-66.

[183] Olsder, G. J. and Walter, J., Collision avoidance of ships, in: Stoer, J. (ed.), Proceedings 8th IFIP Conference on Optimization Techniques, Springer Lecture Note No. 6, in Control and Information Sciences (Springer Verlag, Vienna, 1978), pp. 264-271.

[184] Ozguner, U. and Perkins, W. R., A Nash feedback approach to control of river pollution, IEEE Transactions on Systems, Man and Cybernetics, SMC-9:5 (May 1979).

[185] Pachter, M., Some properties of the value matrix in infinite-time linear-quadratic differential games, IEEE Transactions on Automatic Control, AC-23:4 (August 1978).

[186] Pachter, M. and Getz, W. M., The geometry of the barries in the game of two cars, Optimal Control, Applications and Methods (1980).

[187] Pachter, M. and Yavin, Y., A stochastic homicidal chauffeur pursuit-evasion differential game, Optimal Control Applications and Method, 2:2 (1981) 145-163.

[188] Papavassilopoulos, George P. and Cruz, Jose B., Jr., Nonclassical control problems and Stackelberg games, IEEE Transactions on Automatic Control, AC-24:2 (April 1979) 155-166.

[189] Parthasarathi, T. and Stern, M., On Markov games, in: Roxin et al. (eds.), Differential Games and Control Theory III (Marcel Dekker, New York, 1977) pp. 1-46.

[190] Peng, W. Y. and Vincent, T. L., Some aspects of aerial combat, AIAA Journal 13:1 (January 1975) 7-11.

[191] Petrosyan, L. A., in: Dokl. Akad. Nauk SSSR 195:3 (1970).

[192] Petrosyan, L. A., Topological games and their applications to pursuit games, SIAM Journal of Control 10:1 (1972) 194.

[193] Pindyck, R. S., Optimal economic stabilization policies under decentralized control and conflicting objectives, IEEE Transactions on Automatic Control, AC-22 (1977).

[194] Pontrjagin, L., Linear differential games I, Soviet Math. Doklady 8:3 (1967) 769-771. (a)

[195] Pontrjagin, L., Linear differential games II, Soviet Math. Doklady 8:4 (1967) 910-912. (b)

[196] Pontryaguine, L. S., On the evasion process in differential games, Applied Mathematics and Optimization 1 (1974). (a)

[197] Pontryaguine, L. S., Linear differential games, SIAM Journal of Control 12:2 (May 1974). (b)

[198] Pshenichnyi, B. N., Linear differential games, Automation and Remote Control 29:1 (1968) 55-67.

[199] Ray, A. and Blaquiere, A., Sufficient conditions for optimality of threat strategies in a differential game, Journal of Optimization Theory and Applications (1980).

[200] Rechtschaffen, E. E. M., Unique winning policies for linear differential pursuit games, Journal of Optimization Theory and Applications 29:4 (1979) 629-658.

[201] Rhodes, I. B. and Luenberger, D. G., Differential games with imperfect state information, IEEE Transactions on Automatic Control, AC-14 (1969) 29-38.

[202] Roxin, E. O., The axiomatic approach in differential games, Journal of Optimization Theory and Applications 3 (1969) 153-163.

[203] Roxin, E. O., Differential games with partial differential equations, in: Hagedorn, et al. (eds.), Differential Games and Applications, Lecture Note No. 3 in Control and Information Sciences (Springer Verlag, Vienna, 1977) pp. 186-204.

[204] Sarma, Y. G., and Prasad, U. R., Switching surfaces in N-person differential games, in: Leitmann, G. (ed.), Multicriteria Decision Making and Differential Games (Plenum, 1976), pp. 211-32.

[205] Shapley, L. S., Stochastic games, Proceedings of the National Academy of Sciences, USA 39 (1953) 1095-1100.

[206] Sharma, L. D., On ship maneuverability and collision avoidance, Internal Report Institut fuer Schiffbau der Universitaet Hamburg (1976).

[207] Shubik, M. and Sobel, M. J., Stochastic games, oligopoly theory, and resource allocation, in: Liu P.-T. (ed.), Dynamic Optimization and Mathematical Economics (Plenum, New York, 1980).

[208] Simaan, M. and Cruz, J. B., Jr., On the Stackelberg strategy in nonzero sum games, Journal of Optimization Theory and Applications 11:5 (May 1973) 533-545. (a)

[209] Simaan, M. and Cruz, J. B., Jr., Additional aspects of the Stackelberg strategy in nonzero-sum games, Journal of Optimization Theory and Applications 11:6 (June 1973) 613-626. (b)

[210] Simaan, M. and Cruz, J. B., Jr., Nash equilibrium strategies for the problem of armament race control, Management Science 22:1 (1975) 96-105. (a)

[211] Simaan, M. and Cruz, J. B., Jr., Formulation of Richardson's model of arms race from a differential game viewpoint, The Review of Economic Studies 42:1 (1975) 67-77. (b)

[212] Simaan, M. and Cruz, J. B., Jr., Equilibrium concepts for arms race problems, in: Gillespie, J. V. and Zinnes, D. A. (eds.), Mathematical Systems in International Relations Research (undated), Ch. 11.

[213] Simaan, M. and Takayama, T., Game theory applied to dynamic duopoly problems with production constraints, Automatica 14 (1978) 161-166.

[214] Sobel, M. J., Noncooperative stochastic games, Annals of Math. Statistics 42 (1971) 1930-1935.

[215] Stadler, W., A survey of multicriteria optimization or the vector maximum problem, part I, 1776-1960, Journal of Optimization Theory and Applications 29:1 (1979) 1-52.

[216] Starr, A. W. and Ho, Y. C., Nonzero-sum differential games, Journal of Optimization Theory and Applications 3:3 (1969). (a)

[217] Starr, A. W. and Ho, Y. C., Further properties of nonzero-sum differential games, Journal of Optimization Theory and Applications 3:4 (1969). (b)

[218] Sun, F. K. and Ho, Y. C., Value of information in zero-sum games, Math. Program. Study 5 (1976) 211-226. (a)

[219] Sun, F. K. and Ho, Y. C., Role of information in the stochastic zero-sum differential game, in: Leitman, G. (ed.), Multicriteria Decision Making and Differential Games (Plenum, New York, 1976) Ch. 19. (b)

[220] Sun, F. K. and Ho, Y. C., Note on the linear quadratic Gaussian zero-sum differential game with asymmetric information structures, in: Leitmann, G. (ed.), Multicriteria Decision Making and Differential Games (Plenum, New York, 1976) Ch. 20. (c)

[221] Taylor, J. G., Necessary conditions of optimality for a differential game with bounded state variables, IEEE Transactions on Automatic Control, AC-20 (December 1975) 807-808.

[222] Taylor, J. G., Recent development in the Lanchester theory of combat, in: Haley (ed.), Operational Research (North Holland, Amsterdam, 1978) pp. 773-806.

[223] Tolwinski, B., Closed-loop Stackelberg solution to multistage linear-quadratic game, Journal of Optimization Theory and Applications (1980).

[224] Varaiya, P. and Lin, Y., Existence of saddlepoint in differential games, Journal of SIAM Series A, Control 7 (1969) 141-157.

[225] Vincent, T. L., Avoidance of guided projectiles, in: Grote, J. D. (ed.), The Theory and Applications of Differential Games (Reidel, Dordrecht, Holland, 1975).

[226] Vincent, T.L., Collision avoidance at sea, Lecture Notes in Control and Information Sciences, Vol. 3 (Springer Verlag, Vienna, 1977).

[227] Vincent, T. L., Cliff, E. M., Grantham, W. J. and Peng, W. Y., Aircraft collision avoidance, paper presented at the Workshop on Differential Games, Naval Academy, Annapolis (1973).

[228] Vincent, T. L., Cliff, E. M., Granthan, W. J. and Peng, W. Y., Some aspects of collision avoidance, AIAA Journal 12:1 (January 1974) 3-4.

[229] Vincent, T. L. and Peng, W. Y., Ship collision avoidance, Navy Workshop in Differential Games, Annapolis, July 31-August 3 (1973).

[230] Vincent, T. L., Sticht, D. J. and Peng, W. Y., Aircraft missile avoidance, Journal of Operations Research 24:3 (May-June 1976).

[231] Willman, W. M., Formal solutions for a class of stochastic-pursuit-evasion games, IEEE Transactions on Automatic Control, AC-14 (1969) 504-509.

[232] Wilson, D. J., Isaacs' princess and monster game on the circle, Journal of Optimization Theory and Applications 9 (1972) 265-288.

[233] Wishart, D. M. and Olsder, G. J., Discontinuous Stackelberg solutions, International Journal of Systems Sciences 10:12 (1979) 1359-1368.

[234] Witsenhausen, H. S., On the relations between the values of a game and its information structure, Information and Control 19:2 (1971) 204-215.

[235] Witsenhausen, H. S., On evaluation competitive information, Proc. 1977 Lawrence Symposium on Systems and Decision Science, Berkeley, California (1977).

[236] Worsham, R. H., A discrete game with a mobile hider, in: Roxin, E. O., et al. (eds.), Differential Games and Control Theory (Dekker, New York, 1974).

[237] Zelikin, M. Y., On a differential game with incomplete information, Soviet Math. Dohl. 13 (1972) 228-231.

Other References

[238] Bagchi, A. and Olsder, G. J., A stochastic differential game as control problem with infinite dimensional state space, in: Banks, S. and Pritchard, A. (eds.), Proceedings 2nd IFAC Symposium on Control of Distributed Parameter Systems (Pergamon, Helsinki, 1978).

[239] Barron, E. N., Differential games with Lipschitz control functions and fixed initial control positions, Journal of Differential Equations 26 (1977) 161-180.

[240] Barron, E. N., Remarks on the existence of value in differential games, Journal of Math. Analysis and Applications 58 (1977) 631-636.

[241] Basar, T. and Olsder, G. J., Mixed Stackelberg strategies in continuous kernel games, Reprints CDC Conference, Fort Lauderdale, Florida (1979).

[242] Basile, G. and Vincent, T. L., On the cooperative differential game, Proceedings of the Third Hawaii International Conference on Systems Sciences, Honolulu, Hawaii (January 1970). (a)

[243] Basile, G. and Vincent, T. L., The absolutely cooperative solution for a linear multiplayer differential game, Journal of Optimization Theory and Applications 6:1 (July 1970). (b)

[244] Bensoussan, A., Saddle points of convex concave functionals, in: Kuhn and Szego (eds.), Differential Games and Related Topics (North Holland, Amsterdam, 1971).

[245] Bensoussan, A., Points de Nash dans le cas de fonctionnelles quadratiques et jeux differentiels lineaires a n personnes, SIAM Journal on Control 12 (1974).

[246] Berkovitz, L. D., A variational approach to differential games, Advances in Game Theory, Annals of Mathematics Study #52, M. Dresher, ed., 1964, Princeton University Press.

[247] Berkovitz, L. D., Necessary conditions for optimal strategies in a class of differential games and control problems, Journal SIAM Control 5:1 (1967).

[248] Bernhard, P. and Abramtic, J. F., Contraintes d'etat dans les jeux differentiels, in: Bensoussan, A. and Lions, J. L. (eds.), Control Theory Numerical Methods and Systems Modelling, Lecture Notes in Economics and Mathematical Systems, Vol. 107 (Springer Verlag, Vienna, 1975).

[249] Blaquiere, A., Quantitative games: problem statement and examples, new geometric aspects, in: Gote, J. D. (ed.), The Theory and Application of Differential Games (Reidel, Holland, 1975), pp. 109-120.

[250] Blaquiere, A. and Caussin, P., Further geometric aspects of differential games, in: Blaquiere, A. (ed.), Topics in Differential Games (North Holland, Amsterdam, 1973) pp. 101-149.

[251] Blaquiere, A., Juricek, L., and Wiese, K. E., Geometry of Pareto equilibria and a maximum principle in N-person differential games, Journal of Mathematical Analysis and Applications 38: 1 (April 1972) 223-243.

[252] Blaquiere, A., Juricek, L., and Wiese, K. E., Geometry of Pareto equilibria in N-person differential games, in: Blaquiere, A. (ed.), Topics in Differential Games (North Holland, Amsterdam, 1973) pp. 271-310.

[253] Blaquiere, Austin and Leitmann, George, Multistage quantitative games with unprescribed terminal stages, International Journal of Non-Linear Mechanics 5 (1970).

[254] Borgest, W. and Varaiya, P., Target function approach to linear pursuit problems, IEEE Transactions on Automatic Control, AC-16:5 (October 1971).

[255] Breakwell, J. V. and Merz, A. W., Minimum required capture radius in a coplanar model of the aerial combat problem, AIAA Journal 15:8 (August 1977).

[256] Brunovsky, P., Completing linear differential games by state dependent strategies, Kybernetika 10:1 (1974).

[257] Case, J., A differential game in economics, Management Science 17:7 (March 1971).

[258] Chen, C. I. and Cruz, Jose B., Jr., Stackelberg solution for two-person games with biased information patterns, IEEE Transactions on Automatic Control, AC-17:6 (December 1972) 791-798.

[259] Chen, Santiago Fei-Hung and Leitmann, George, Labor-management bargaining modeled as a dynamic game, Journal of Optimization Control Application and Methods (1980).

[260] Chentsov, A. G., An example of an irregular differential game, PMM 40:6 (1976) 1113-1116.

[261] Chikri, A. A. and Chikri, G.Ts., On a differential-difference game of escape, PMM 40:6 (1976) 995-1002.

[262] Chyung, D. H., A pursuit-evasion game of linear systems with delay, IEEE Transactions on Automatic Control (June 1970).

[263] Clemhout, S., Leitmann, G. and Wan, H. Y., Jr., A differential game model of duopoly, Econometrica 39:6 (November 1971) 911-935.

[264] Cruz, J. B., Jr. and Chen, C. I., Series Nash solution of two-person, nonzero-sum, linear-quadratic differential games, Journal of Optimization Theory and Applications 7:4 (April 1971) 240-257.

[265] Dobbie, J. M., Solution of some surveillance evasion problems by the methods of differential games, 4th International Conference on Operational Research, M.I.T. (1966).

[266] Dolezal, J. and Thoma, M., On the numerical solution of N-player nonzero-sum differential games via Nash solution concept, Problems of Control and Information Theory 8:5-6 (1979) 371-385.

[267] Duchateau, C., Statistical linear differential games: a treatment of incomplete information, Journal of Optimization Theory and Applications 19:2 (June 1976).

[268] Elliott, R. J. and Kalton, N. J., Upper values of differential games, Journal of Differential Equations 14 (1973) 89-100.

[269] Elliott, R. J., Kalton, N. J. and Markus, L., Saddle points for linear differential games, SIAM Journal Control 11:1 (February 1973).

[270] Foley, M. H. and Schmitendorf, W. H., On a class of nonzero-sum, linear-quadratic differential games, Journal of Optimization Theory and Applications 7:5 (1971).

[271] Friedman, A., Stochastic differential games, Journal of Differential Equations 11 (1972) 79-108.

[272] Friedland, S., Global principle for free-endpoint problems in optimal control and differential games, Journal of Optimization Theory and Applications 24:2 (February 1978).

[273] Gamkrelidze, R. V., On certain integral equations related to differential games, Banach Center Publication, Vol. 1 (undated) 83-88, in Russian.

[274] Gamkrelidze, R. V. and Khararishvili, G. L., A differential game of evasion with nonlinear control, SIAM Journal Control 12:2 (May 1974).

[275] Gardner, B. F., Jr. and Cruz, J. B., Jr., Feedback Stackelberg strategy for a two player game, IEEE Transactions on Automatic Control (April 1977) 270-271.

[276] Gardner, B. F., Jr. and Cruz, J. B., Jr., Well-posedness of singularly perturbed Nash games, Journal of the Franklin Institute 306:5 (November 1978) 355-374.

[277] Gorokhovik, V. V. and Kirillova, F. M., About many players linear differential games, Control Systems, IM, IK SO AN SSSR, issue 10, Novosibirsk (1971) 3-9.

[278] Gutman, S. and Leitmann, G., On a class of linear differential games, Journal of Optimization Theory and Applications 17:5-6 (December 1975) 511-522.

[279] Gutman, S. and Leitmann, G., Optimal strategies in the neighborhood of a collision course, AIAA Journal 14:9 (September 1976) 1210-1212.

[280] Heymann, M., Pachter, M. and Stern, R. J., On linear games with subspace target, Funkcialaj Ekvacioj 20:1 (April 1977). (a)

[281] Heymann, M., Pachter, M. and Stern, R. J., Two-player control problems with subspace targets, in: Roxin et al. (eds.), Differential Games and Control Theory (Dekker, New York, 1977). (b)

[282] Ho, Y. C. and Baron, S., A minimal time intercept problem, IEEE Transactions on Automatic Control (April 1965).

[283] Isaacs, R., Some fundamentals on differential games, in: Blaquiere, A. (ed.), Topics in Differential Games (North Holland, Amsterdam, 1973).

[284] Jumaire, G., Stability, structural stability and games: toward a theory of differential hypergames, Journal of Optimization Theory and Applications 22:4 (August 1977).

[285] Kaskosz, B., On a nonlinear evasion problem, SIAM Control and Optimization 15:4 (July 1977).

[286] Kelley, Henry J., Differential-turn maneuvering, Automatics 12 (1976) 257-259. (a)

[287] Kelley, H. J., A threat-reciprocity concept for pursuit/evasion, presented at the IFAC Differential Games Conference, University of Rhode Island, Kingston, R.I., June 7-10 (1976). (b)

[288] Kirillova, F. M. and Kirov, S. M., The application of functional analysis to problems of pursuit, in: Balakrishnan, A. V. and Neustadt, Lucien W. (eds.), Mathematical Theory of Control, Proceedings of a Conference held at the University of Southern California, Los Angeles, January 30-February 1 (Academic Press, 1967), pp. 386-397.

[289] Kloeden, P. E., An attainability set model for dynamical games, Journal of Differential Equations 21 (1976) 197-230.

[290] Korneev, V. A. and Melikian, A. A., Differential game with impulsive control of one of the players, PMM 41:2 (1977) 354-357.

[291] Krasovskii, N. N., On the problem of pursuit in the case of linear monotype objects, PMM 30:2 (1966) 209-225.

[292] Krasovskii, N. N., On capture in differential games, Soviet Math. Dokl. 9:4 (1968).

[293] Krasovskii, N. N., External control in a nonlinear differential game, PMM 36:6 (1972) 986-1006.

[294] Krasovskii, N. N., External constructions for a differential game, Soviet Math. Dokl. 11:1 (1977).

[295] Krasovskii, N. N. and Subbotin, A. I., On the structure of differential games, Soviet Math. Dokl. 11:1 (1970).

[296] Lakshmikantham, V. and Rao, A. N. V., Differential inequalities and differential games, Journal of Math. Analysis and Applications 41 (1973) 490-496.

[297] Leitmann, George, A simple differential game, Journal of Optimization Theory and Applications 2:4 (1968) 220.

[298] Leitmann, George, Differential game of pursuit and evasion, International Journal of Nonlinear Mechanics 4:1 (1969).

[299] Leitmann, George, Cooperative and non-cooperative differential games, in: Gote, J. D. (ed.), The Theory and Application of Differential Games (Reidel, Amsterdam, 1975), pp. 85-96.

[300] Leitmann, George, Optimal feedback control for dynamical systems with one or two controllers, in: Mathematical Theory of Dynamical Systems and Microphysics (Springer Verlag, Vienna, 1980).

[301] Leitmann, George and Blaquiere, A., Multistage quantitative games, Proc. Int. Col. on Methods of Optimization, Novosibirsk, 1968 (Springer Verlag, Vienna, 1970).

[302] Leitmann, George, Clemhout, S. and Wan, H., A differential game model of duopoly, Proc. 1st Int. Conf. on Theory and Appl. on Diff. Games, II-11 (1969).

[303] Leitmann, George and Liu, P. T., A differential game model of labor-management negotiation during a strike, 5th IFIP Conf. on Optimization Techniques, Journal of Optimization Theory and Applications 13:4 (1972) 427 and 14:4 (1974) 443.

[304] Leitmann, George and Liu, H. S., Evasion in the plane, Proceedings of 8th IFIP Conf. on Optimization Techniques, Springer Verlag, Berlin (1977), pp. 255-263.

[305] Leitmann, George and Schmitendorf, W., Profit maximization through advertising: a nonzero sum differential game approach, IEEE Transactions, AC-23:4 (1978) 645.

[306] Leitmann, G., Rocklin, S. and Vincent, T. L., A note on control space properties of cooperative games, Journal of Optimization Theory and Applications 9:6 (1972).

[307] Leitmann, George and Skworonski, J., Avoidance control, Journal of Optimization Theory and Applications 23:4 (1977) 581.

[308] Leitmann, George and Stalford, H. L., Sufficient conditions for optimality in two-person zero-sum differential games with state and strategy constraints, Journal Math. Anal. Appl. 33 (1971) 650.

[309] Leitmann, George and Stalford, H., Sufficiency conditions for Nash equilibria in N-person differential games, in: Blaquiere, A (ed.), Topics in Differential Games (North-Holland, Amsterdam, 1973).

[310] Leitmann, George and Stalford, H., On saddle-point optimality in differential games, Journal of Optimization Theory and Applications 18:1 (1976) 141.

[311] Levine, J., Zero sum two person differential games with incomplete information: a Bayesian approach, 3rd Kingston Conference on Differential Games and Control Theory (1978).

[312] Liu, P. T., On a problem of differential games with imperfect information, Proc. of the First International Conference on Differential Games (September 1969).

[313] Liu, P. T., Linear pursuit-evasion games with bounded state spaces, Journal of Optimization Theory and Applications 8:3 (1971). (a)

[314] Liu, P. T., A minimax problem for distributed parameter systems, International Journal of Control 13:5 (1971). (b)

[315] Liu, P. T., Linear pursuit-evasion games with bounded state spaces, Journal of Optimization Theory and Applications 8:3 (September 1971). (c)

[316] Liu, P. T., Pursuit-evasion games with genralized terminal conditions, Journal of Math. Anal. and Appl. 42:2 (1973) 499-506.

[317] Liu, P. T., A game theoretic approach to optimum speeds in antisubmarine warfare, Technical Report, Naval Underwater Systems Center, Newport, R.I. (September 1976).

[318] Liu, P. T., Stochastic pursuit-evasion games, Journal of Math. Anal. and Appl. 59:2 (June 1977).

[319] Liu, P. T., Pareto optimal leasing and investment policies in the exploitation of exhaustible natural resource, in: Aoki and Marzollo (eds.), New Trends in Dynamic System Theory and Economics (Academic Press, New York, 1979).

[320] Mageirou, E. F. and Ho, Y. C., Decentralized stabilization via game theoretic methods, Automatica 13 (1977) 393-399.

[321] Maksimov, V. I., Linear differential encounter game with a functional target, PMM 40:2 (1976) 222-229.

[322] Maksimov, V. I., On the existence of a saddle-point in a difference-differential encounter-evasion game, PMM 42:1 (1978) 15-22.

[323] Matsumoto, T., On some properties of min-max functions, SIAM Journal on Control and Optimization 14:1 (January 1976).

[324] Matsumoto, T. and Shimemura, E., On a class of linear pursuit games, Journal of Differential Equations 11:2 (March 1972).

[325] Men'shikov, I.S., Hierarchical differential-game, PMM 42:1 (1978) 23-27.

[326] Meschler, P. A., On a goal-keeping differential game, IEEE Transactions on Automatic Control, AC-12:1 (February 1967).

[327] Mischenko, E. F., On the problem of evading the encounter in differential games, SIAM Journal Control 12:2 (May 1974).

[328] Olsder, G. J., Comment on a numerical procedure for the solution of differential games, IEEE Transactions on Automatic Control, AC-20:5 (1975) 704-705.

[329] Papavassilopoulos, G. P. and Cruz, J. B., Jr., On the existence of solutions to coupled matrix Riccati differential equations in linear quadratic Nash games, IEEE Transactions on Automatic Control, AC-24:1 (February 1979) 127-129. (a)

[330] Papavassilopoulos, G. P. and Cruz, J. B., Jr., On the uniqueness of Nash strategies for a class of analytic differential games, Technical Note, Journal of Optimization Theory and Applications 27:2 (1979) 309-314. (b)

[331] Papavassilopoulos, G. P., Medanic, J. V. and Cruz, J. B., Jr., On the existence of Nash strategies and solutions to coupled Riccati equations in linear-quadratic games, Journal of Optimization Theory and Applications 28:1 (May 1979) 49-76.

[332] Ragade, R. K. and Sarma, I. G., A note on controllability, observability and sufficient coordinates in linear differential games, International Journal Control 10:3 (1969) 331-335.

[333] Ray, A., Optimal threat strategies in collective bargaining, Internal Report, Laboratoire d'Automatique Theorique, Universite de Paris 7, Paris, France (1979).

[334] Reshetov, V. M. and Ushakov, V. N., Stabilization in a linear differential game with integral constraints on the players' control resources, PMM 40:3 (1976) 439-445.

[335] Roxin, E. O., Some global problems in differential games global differentiable dynamics, in: Hajek, O., Lowater, A. L. and McCann, R. (eds.), Lecture Notes in Math No. 235 (Springer Verlag, Vienna, 1971).

[336] Roxin, E., Axiomatic approach in differential games, Journal of Optimization Theory and Applications 3:3 (1969).

[337] Rubinovich, E. Ya, Choosing preferable strategies in a simple differential game of piecewise-programmed pursuit, Avtomatika Telemekhanika No. 10 (October 1978), pp. 48-54.

[338] Rublein, G. T., On pursuit with curvature constraints, SIAM J.Control 10:1 (February 1972).

[339] Salman, M. A. and Cruz, J. B., Jr., Well-posedness of linear closed-loop Stackelberg strategies for singularly perturbed systems, Journal of the Franklin Institute 308:1 (July 1979) 25-37.

[340] Schmitendorf, W. E., Differential games with open-loop saddle point conditions, IEEE Transactions on Automatic Control, AC-15:3 (June 1970).

[341] Schmitendorf, W. E., Cooperative games and vector-valued criteria problems, IEEE Transactions on Automatic Control, AC-18:2 (April 1973).

[342] Schmitendorf, W. E., Pontryagin's principle for problems with isoperimetric constraints and for problems with inequality terminal constraints, Journal of Optimization Theory and Applications 18:4 (April 1976).

[343] Shinar, J. and Steinberg, D., Analysis of optimal evasive maneuvers based on a linearized-two-dimensional kinematic model, Journal of Aircraft 14:8 (August 1977).

[344] Skowronski, J. M. and Vincent, T. L., Playability with and without capture, University of Queensland Control Theory Report 77-3 (March 1977).

[345] Sticht, J., Vincent, T. L. and Schultz, D. G., sufficiency theorems for target capture, Journal of Optimization Theory and Applications 17:3/4 (November 1975).

[346] Sun, Fang-Kuo and Ho, Yu-Chi, Value of information in zero-sum games, Math. Programming Study 5 (1976) 211-226.

[347] Tomskii, G. V., Encounter-evasion problems in quasidynamic systems, PMM 42:2 (1978) 219-227.

[348] Varaiya, P. P., On the existence of solutions to a differential game, SIAM J. Control 5:1 (1967).

[349] Varaiya, P., N-person non-zero differential games with linear dynamics, SIAM J. Control 8:4 (November 1970).

[350] Varaiya, P., N-player stochastic differential games, SIAM J. Control and Optimization 14:3 (May 1976).

[351] Vidyasagar, M., A new approach to N-person, nonzero-sum, linear differential games, Journal of Optimization Theory and Applications 18:1 (January 1976).

[352] Vincent, T. L., Game theoretic predictions of genetic shifts in a plant breeding model, Proceedings of the Twenty-First Midwest Symposium on Circuits and Systems, Iowa State University, Ames, August 14-15 (1978).

[353] Vincent, T. L., Cliff, E. M., and Grantham, W. J., On barriers in a problem of collision avoidance, Proceedings of the Fourth Hawaii International Conference on System Sciences, Honolulu, Hawaii, January (1971).

[354] Vincent, T. L., Cliff, E. M., Grantham, W. J. and Peng, W. Y., A problem of collision avoidance, University of Arizona, Engineering Experiment Station, Report 39, November (1972).

[355] Vincent, T. L. and Peng, W. Y., A problem of aerial combat, Navy Workshop in Differential Games, Annapolis, Maryland, July 31-August 3 (1973).

[356] Walsh, P. M. and Cruz, J. B., Jr., A sampled data equilibrium Stackelberg coordination scheme for the multicontroller problem, IEEE Transactions on Automatic Control, AC-23:4 (August 1978).

[357] Wilson, D. J., Differential games with no information, SIAM J. Control and Optimization 15:2 (February 1977).

[358] Witsenhausen, H. S., A minimax control problem for sampled linear systems, IEEE Transactions on Automatic Control, AC-13:1 (February 1968).